Foreword

This is the Project Report for CIRIA Research Project 492 *Inland dredging techniques and operations - guidance on good practice*, produced as part of CIRIA's water engineering programme. It also constitutes Environment Agency R&D Project Record i588/1/W5. The primary objective of the project was to prepare guidance on good practice for inland dredging and sediment clearance operations, including consideration of costs, environmental effects and contractual obligations.

The report was written under contract to CIRIA by Anthony Bates, Anthony D Bates Partnership, and Alan Hooper, WS Atkins Consultants Ltd. CIRIA's Research Manager for the project was Sian John, with support from Max Wheeler.

Steering Group

Following established CIRIA practice the research was guided by a Steering Group which comprised:

Chairman

Nick Bray	Dredging Research Ltd.

Members

Peter Barham	Environment Agency
David Bligh	British Waterways
John Buckingham	Environment Agency
Kieran Conlan (from spring 1996)	Consultants in Environmental Sciences Ltd.
Iain Fairgrieve	Federation of Dredging Contractors
Charles Ford	Charles R Ford & Associates
Stewart Kirkwood	The Rivers Agency, Department of Agriculture Northern Ireland
John Laverick	The Broads Authority
Simon Mowat	Land and Water Services Ltd.
Topsy Rudd (to winter 1995)	Consultants in Environmental Sciences Ltd.

Corresponding Members

Jan Brooke	Posford Duvivier Environment
Mary Gibson	English Nature
Roger Hanbury	British Waterways
Ray Howells	Manchester Ship Canal Company
David Noble	Association of Drainage Authorities
Julie Lyons	ICI Chemicals & Polymers Ltd.

Project funders

The project was funded by:

British Waterways
The Rivers Agency, Department of Agriculture Northern Ireland (DANI)
National Rivers Authority (NRA)[1]
ICI Chemicals and Polymers Ltd.
The Broads Authority

Acknowledgements

CIRIA is grateful for the support given to the project by the funders, the members of the Steering Group and all of the organisations and individuals who responded to the project questionnaire or participated in the consultation process. In particular:

Association of Drainage Authorities (ADA)
British Waterways - Gloucester
British Waterways - Leeds
British Waterways - Southern
British Waterways - Tamworth
Conbar Consultants Ltd.
De Boer BV
Department of Agriculture Northern Ireland
EMCC, Paris
English Nature
Land and Water Services Ltd.
Manchester Ship Canal Company
May Gurney Ltd.
Ministry of Works - Waterways, Gent
National Rivers Authority
Ponds Group, Oxford
Rijkswaterstaat
Shepway District Council
Silt NV, Antwerp
The Broads Authority
University of Liverpool
Water Group Ltd.

---

[1] As from April 1996, the functions of the National Rivers Authority, Her Majesty's Inspectorate of Pollution and the Waste Regulation Authorities, were taken over in England and Wales by the Environment Agency. Its equivalent functions in Scotland were taken over by the Scottish Environmental Protection Agency (SEPA), and in Northern Ireland by the Environment and Heritage Service.

Report 169   1997

Inland Dredging — guidance on

CIRIA
CONSTRUCTION INDUSTRY RESEARCH AND INFORMATION ASSOCIATION
6 Storey's Gate, Westminster, London SW1P 3AU
E-mail switchboard @ ciria.org.uk
Tel 0171-222 8891   Fax 0171-222 1708

Summary

Waterway related organisations identified inland dredging as an area with significant scope for improvement in current practice. They recognised that considerable savings could be achieved by improving operational effectiveness and efficiency. Inland dredging operations account for millions of pounds of spending in the UK annually. Hence there was a clear need for research to review current technologies, establish operational and environmental good practice, clarify contractual obligations, and to propose a way forward.

CIRIA Report 169 has prepared guidance on good practice with regard to dredging and sediment treatment techniques, including consideration of their utility and flexibility, costs and environmental effects, and in the broader sense on the management of dredging operations. It identifies good practice, defines an appropriate approach to project evaluation, task definition and option selection, considers measurement techniques, contractual relationships and relevant health and safety issues, and outlines further research needs.

In parallel, CIRIA Report 157 *Guidance on the disposal of dredged material to land* presents guidance (consistent with the aims of the Environmental Protection Act 1990) on the safe, economic and effective disposal of dredged material. It identifies how Government advice applies to dredgings, describes how control is to be exercised over disposal, and assists operators in interpreting and operating within the Waste Management Licensing Regulations 1994.

Together, these documents aim to ensure that a framework is established for dredging operations to be undertaken cost effectively and within the law.

Inland dredging - guidance on good practice
Construction Industry Research and Information Association
Report 169, 1997
© CIRIA 1997
ISBN: 0 86017 477 8
ISSN: 0305 408X

Keywords		
Broads, dredger, dredging operations, canals, inland dredging, lakes, maintenance, management, navigation, ponds, reservoirs, rivers, waterway environment, waterways		
Reader interest	**Classification**	
Organisations with responsibility for dredging operations, dredging consultants and contractors, and environmental groups	AVAILABILITY CONTENT STATUS USER	Restricted Guidance on inland dredging techniques and operations Committee guided Dredging operators, clients and regulators

Issued by CIRIA 6 Storey's Gate, Westminster, London SW1P 3AU. All rights reserved. No part of this publication may be reproduced or transmitted in any form or by any means, including photocopying and recording, without the written permission of the copyright holder, application for which should be addressed to the publisher. Such written permission must also be obtained before any part of this publication is stored in a retrieval system of any nature.

Contents

List of tables		9
List of figures		9
List of boxes		10

GLOSSARY		11
1	INTRODUCTION	15
	1.1 BACKGROUND	15
	1.2 OBJECTIVES	15
	1.3 SCOPE AND APPROACH	16
	1.4 READERS' GUIDE	16
	1.5 COMPLIMENTARY PUBLICATIONS	17
2	CURRENT PRACTICE	19
	2.1 INTRODUCTION	19
	2.2 OVERVIEW OF THE NATURE AND SCALE OF OPERATIONS IN ENGLAND, IRELAND AND WALES	19
	2.3 OVERVIEW OF CURRENT PLANNING OF WORK	20
	2.4 GENERAL APPLICATION OF PRIMARY LAND-BASED EXCAVATION EQUIPMENT	21
	2.5 GENERAL APPLICATION OF PRIMARY FLOATING DREDGING EQUIPMENT	22
	2.6 TREATMENT METHODS	23
	2.7 DISPOSAL AND TRANSPORT OF DREDGED MATERIAL	25
	2.7.1 Disposal	25
	2.7.2 Transport	26
	2.8 PRINCIPAL PROBLEM AREAS	27
	2.9 CONTINENTAL EXPERIENCE	30
3	PROJECT EVALUATION AND IMPLEMENTATION	35
	3.1 INTRODUCTION	35
	3.2 OBJECTIVE - REQUIRED LEVEL OF SERVICE	37
	3.3 ALTERNATIVES TO DREDGING	37
	3.4 PROJECT PLANNING	38
	3.4.1 Introduction	38
	3.4.2 Frequency and extent of dredging	39
	3.4.3 Canals and navigable waterways	40
	3.4.4 Rivers and watercourses	41
	3.4.5 Ponds, lakes and reservoirs	41
	3.5 DEFINITION OF THE TASK	42
	3.6 ENVIRONMENTAL CONSIDERATIONS	42
	3.7 ARCHAEOLOGY	44
	3.8 DREDGING, TREATMENT AND DISPOSAL	45
	3.9 CONTRACT VERSUS DIRECT LABOUR	45
	3.10 SPECIALIST INPUT AND TRAINING	46
	3.11 SELECTION OF AN APPROPRIATE DREDGING METHOD	47

4	TASK DEFINITION		55
	4.1	ACCESS	55
		4.1.1 Introduction	55
		4.1.2 Overland	57
		4.1.3 Floating	61
	4.2	SEDIMENT CHARACTERISTICS	67
		4.2.1 Physical characteristics	67
		4.2.2 Chemical and biological characteristics	68
		4.2.3 Pollution by debris	69
	4.3	VEGETATION	71
	4.4	ENVIRONMENTAL CONSIDERATIONS	72
	4.5	SEASONAL FACTORS	73
	4.6	DISPOSAL OPTIONS	74
	4.7	TREATMENT NEEDS	77
	4.8	DIMENSIONS OF THE WORK	77
	4.9	QUANTITIES	79
	4.10	SECURITY	80

5	DREDGING TECHNIQUES		81
	5.1	DRAGLINES	81
		5.1.1 Mechanical characteristics	82
		5.1.2 Operational effectiveness	82
		5.1.3 Mechanical maintenance	83
		5.1.4 Environmental impact	84
	5.2	GRAB DREDGERS	84
		5.2.1 Mechanical characteristics	84
		5.2.2 Operational effectiveness	86
		5.2.3 Mechanical maintenance	86
		5.2.4 Environmental impact	86
	5.3	HYDRAULIC BACKHOES	87
		5.3.1 Mechanical characteristics	88
		5.3.2 Operational effectiveness	89
		5.3.3 Mechanical maintenance	91
		5.3.4 Environmental impact	92
	5.4	PLOUGHS, RAKES AND WATER INJECTION	93
		5.4.1 Environmental impact	95
	5.5	BUCKET DREDGERS	96
		5.5.1 Mechanical characteristics	97
		5.5.2 Operational effectiveness	98
		5.5.3 Mechanical maintenance	98
		5.5.4 Environmental impact	99
	5.6	CUTTER-SUCTION DREDGERS	99
		5.6.1 Mechanical characteristics	100
		5.6.2 Operational effectiveness	101
		5.6.3 Mechanical maintenance	102
		5.6.4 Environmental impact	103
	5.7	WINCH - ENGINES	104
		5.7.1 Mechanical characteristics	104
		5.7.2 Operational effectiveness	104
		5.7.3 Mechanical maintenance	104
		5.7.4 Environmental impact	105
	5.8	LOW TURBIDITY DREDGING	105

	5.9	SEDIMENT TRANSPORT METHODS AND EQUIPMENT	106
		5.9.1 Overland	106
		5.9.2 Road	106
		5.9.3 Hopper barge	107
		5.9.4 Pipeline	108
	5.10	SOIL MODIFICATION	113
	5.11	RELATIVE COSTS	114
	5.12	RELATIVE ENVIRONMENTAL IMPACT	116
6	TREATMENT TECHNIQUES		119
	6.1	ROUTINE TREATMENT TECHNIQUES	119
		6.1.1 Separation	119
		6.1.2 De-watering	121
	6.2	SPECIAL TREATMENT TECHNIQUES	124
		6.2.1 Washing	124
		6.2.2 De-contamination	124
7	CONTRACTUAL MATTERS		127
	7.1	FORM OF CONTRACT	127
	7.2	SPECIFICATION	128
	7.3	METHOD OF MEASUREMENT	129
		7.3.1 Measurement for payment under contract	130
		7.3.2 Survey methods for measurement	132
	7.4	OBLIGATIONS, LEGAL AND CONTRACTUAL	135
	7.5	CONTRACTOR SELECTION, TENDER APPRAISAL, CONTRACT AWARD	137
	7.6	CONTRACT SUPERVISION	139
	7.7	RESPONSIBILITIES	140
8	HEALTH AND SAFETY		143
	8.1	STATUTORY INSTRUMENTS	143
	8.2	RESPONSIBILITIES	144
	8.3	IMPLEMENTATION	144
	8.4	TYPICAL HAZARDS	145
9	ENVIRONMENTAL CONSIDERATIONS		149
	9.1	INTRODUCTION	148
	9.2	CONTROLLING LEGISLATION	149
	9.3	OPERATIONS PLANNING	151
		9.3.1 Identification of key features	152
		9.3.2 Selection of dredging method	152
		9.3.3 Planning of operations	156
	9.4	MONITORING AND AUDITING	165
	9.5	TRAINING	166

10	RECOMMENDATIONS		169
	10.1	MONITORING AND DATA ANALYSIS	168
	10.2	CONSULTATION	169
	10.3	FORM OF CONTRACT	169
	10.4	SELECTION OF CONTRACTOR	170
	10.5	METHOD OF MEASUREMENT	172
	10.6	SCOPE FOR IMPROVEMENT IN LAND BASED METHODS	171
	10.7	SCOPE FOR IMPROVEMENT IN FLOATING METHODS	171
	10.8	FUTURE RESEARCH NEEDS	173

APPENDIX - PUMPING		174
A.1	PUMP HEAD	175
A.2	PUMP POWER	176
A.3	SETTLING TIME OF FINE PARTICLES	177

REFERENCES — 178

List of tables

2.1	Summary of statistics collected on the nature and scale of dredging	20
2.2	Land-based plant - percentages of total number of dredging plant employed	22
2.3	Floating plant - percentages of total number of dredging plant employed	23
2.4	Analysis of transport methods used for the disposal of dredged material	27
3.1	Guide to typical dredging plant capabilities	48
4.1	Typical lifting capacities for modern mobile telescopic cranes	66
5.1	Guide to containment area total capacity	110
5.2	Relative environmental impact of different dredging and transport methods	117
7.1	Working tolerances for inland dredgers	131
9.1	Relative impact assessment - Example 1	155
9.2	Relative impact assessment - Example 2	156
A.1	Typical friction factors for steel pipes	174
A.2	Equivalent pipe lengths in metres	175
A.3	Typical bulk densities for various hydro-soils	176
A.4	Typical velocities appropriate for pumped mixtures	176
A.5	Typical volumetric concentrations for pumped soils	176
A.6	Settling velocity for soil particles	177

List of figures

2.1	Amphibious dredger working in bog conditions	24
2.2	Limited bank access	28
2.3	Accident resulting from bank edge failure	29
3.1	Project implementation procedure	36
3.2	Prediction of dredging need	39
3.3	Containment of suspended solids	44
3.4	Chart 1 - Disposal	49
3.5	Chart 2 - Dimensions and access	50
3.6	Chart 3 - Material	51
3.7	Chart 4 - Environment	52
3.8	Chart 5 - Selection summary	53
4.1	Limited tow path access	56
4.2	Access restrictions in tunnels	56
4.3	Low bridge	57
4.4	Weak ground conditions	59
4.5	Restricted width	60
4.6	Weak access bridge	61
4.7	Obstruction to navigation caused by dredger and hopper barge	63
4.8	Tandem crane lift	66
4.9	Assessment of debris in drained channel	70
4.10	Assorted debris recovered from inland waterway	71
4.11	Bank-side disposal	76
4.12	Disposal on agricultural land	76
4.13	Tracked machine working from bed in shallow water	78

5.1	Operation of dragline	81
5.2	Dragline excavator	82
5.3	Dragline bucket	83
5.4	Different types of grab bucket	85
5.5	Small floating grab dredger	85
5.6	Floating backhoe dredger	88
5.7	Long-reach backhoe	89
5.8	Amphibious dredger	90
5.9	Backhoe mounted on spudded pontoon	90
5.10	Long-reach machine with mixed hydraulic and rope operation	91
5.11	Weed removal attachment (rake)	92
5.12	Use of a backhoe fitted with a rake to remove debris	93
5.13	Workboat and plough	94
5.14	Simple water injection equipment	95
5.15	Schematic of a bucket dredger	97
5.16	Mooring pattern and method of operation of a bucket dredger	98
5.17	Schematic of a cutter-suction dredger	100
5.18	Crown and auger type cutter heads	101
5.19	Discharge of hopper barge by land based backhoe	108
5.20	Transport by pipeline	109
5.21	Suction dredger pumping to remote containment areas	110
5.22	Extensive containment areas	111
5.23	Weir box	112
6.1	Static screen separator	120
6.2	Powered screen separator	121
6.3	Wet material contained within a shallow bunded enclosure	123
6.4	Shallow bunded enclosure after drainage and evaporation	123
7.1	Laborious survey techniques	133
9.1	Barren artificial channel	148
9.2	Asymmetrical cross-sections	157
9.3	Retention of reed margin	159
9.4	Shaded environment	160
9.5	Intermittent dredging programme	161
9.6	River, pools and riffle system	162
9.7	Example of retained vegetation	163

List of boxes

3.1	Relevant merits of maximum and minimum intervals between dredging	40
3.2	Environmental considerations - list of potential consultees	43
4.1	Principal factors which define the dredging task	55
4.2	Disposal methods listed approximately in ascending order of cost and environmental impact	75
7.1	General specification	128
7.2	Particular specification	129
7.3	Check list for levelling or dipping surveys	134
7.4	Duties of the Employer	140
7.5	Duties of the Contractor	141
7.6	Matters of joint responsibility	141
9.1	Major environmental consequences of dredging	149

Glossary

This report concerns dredging and associated operations. Particular terms used are defined as follows:

Agitation dredging	the practice of forcing sediment from the bed into suspension in the water column for dispersal by local currents.
Backwater	areas of low velocity or stagnant water.
Batter	slope of a bank; if measured the ratio of horizontal to vertical.
Bed	ground at the bottom of the water column in any mass of water.
Blockage factor	the ratio of the submerged cross sectional area of a vessel navigating to the submerged cross sectional area of the waterway; hence the blockage factor is always less than unity.
Bollard	fixture on shore or vessel for attachment of mooring line.
Broad	shallow man made lake of medieval origin in Norfolk and Suffolk.
Bucket capacity	the maximum volume of liquid that can be contained in the bucket when filled to the level of the cutting edge, or enclosing structure.
Bulking factor	a factor representing the increase in volume of dredged material relative to its volume before dredging.
Capital dredging	dredging to increase one or more dimension of a body of water beyond that which had previously existed, i.e. new work or expansion of existing work.
Colonisation	successful occupation of a new habitat by a species not previously present.
Conveyance	flow capacity of a channel.
Cycle time	time to complete a repetitive sequence of dredging actions.
Dredger (or dredge)	mechanical, electrical, or hydraulic plant used for dredging.

Dredging	the removal beneath water of soil, rock or debris.
ESA	Environmentally Sensitive Area (where the landscape, wildlife and historic interest are of national importance).
Faggots	bundles of long branches.
Fairlead	roller, pulley, or fixture to guide winch wire, or rope.
Fendering	protection to side of vessel or structure to resist impact damage.
Freeboard	height above water level of side of vessel, or top of retaining bank.
Geotextiles	natural or synthetic, permeable fabrics used in conjunction with soil to strengthen or protect against erosion.
GIS	Geographical Information System. System, usually computer based, to store, correlate and display geographical information, such as maps and geographical data.
Haul distance	the one way distance that a vessel or vehicle has to travel to a disposal area.
Hydro-cyclone	system for separating coarse and fine soil particles within a wet mixture by centrifugal force.
In-situ density	the unit mass of bottom materials in their undisturbed state.
Jib	of crane or dragline. Long fabricated steel structure to support excavating bucket, or crane hook, at an extended radius.
Levelling staff	hand held vertical staff graduated in metres and 0.01 m for use in conjunction with an optical survey level for levelling over land.
Magnetic separator	type of equipment for treatment process used to separate ferrous metals.
Maintenance dredging	dredging to restore or maintain a depth of water that existed previously, but has been reduced by the deposition of sediment, or debris.
Off-side	in canals, the opposite side to tow path.
Over-dredging	the dredging of material from levels below that specified or required.

Pond	enclosed inland water with a surface area of one acre or less.
Riffle-pool sequence	the system of bed forms found in gravel bed rivers associated with minimisation of energy expenditure whilst permitting sediment transport.
Section	the cross sectional area of a waterway as bounded by the water surface and submerged bed profile.
Scoping study	initial review of available information to establish the need and framework for more detailed studies.
Shoaling	reduction in water depth caused by local accumulation of sediment.
Siltation	the process of deposition of sediments in water.
Skimmers	type of equipment for treatment process used to skim off the surface layer of dredgings, such as floating organic material.
Spud	adjustable vertical steel column used to anchor and locate dredger, or pontoon.
SSSI	Site of Special Scientific Interest.
SPA	Special Protection Area.
SAC	Special Areas for Conservation.
Screening	separation of material of different sizes by passing through or retention on a grid of appropriate size.
Supernatant	upper surface layer of clear water in a reclamation or disposal area.
Tow path	access adjacent to one side of canal originally used by horses towing barges.
Trommel	a rotary screen.
Winding holes	wide section in canal for turning boats.

1 Introduction

Dredging inland waters may be necessary for many reasons. Most commonly it will be necessary to preserve or restore the efficient function of a waterway, whether that function is to permit navigation, or to carry a particular flow of water. In ponds, lakes and reservoirs, dredging may be necessary simply to increase water depth or capacity. Water depth may be important for visual effect and amenity. Capacity may be important for storage. Regardless of the objective, before dredging is undertaken consideration should be given to the need and justification for dredging and the possibility of alternative solutions, including doing nothing. When the need for dredging is established, work should be preceded by appropriate planning to optimise the efficiency of the operation and, most importantly, to minimise the impact of the work on the environment. Dredging need not have an adverse environmental impact, indeed, if a sensitive approach is adopted, the opposite effect is possible.

The purpose of this report is to aid planners and operators in achieving efficient and environmentally friendly solutions to dredging problems through the pursuit and implementation of good practice.

1.1 BACKGROUND

CIRIA Report 169 is complementary to CIRIA Report 157, *Guidance on the disposal of dredged material to land* (CIRIA, 1996a), produced in response to the introduction of the Waste Management Licensing Regulations (DoE, 1994) which came into force on 1 May 1994. Amongst other categories of waste, these Regulations control the disposal of dredged material to land. Under the Regulations, unless it can be shown that dredged material is not waste (i.e. it has not been discarded), or that the proposed disposal method meets one of the specified exemptions, a Waste Management Licence will be required. This requirement remains valid despite the fact that the disposal of dredged material is exempt from the Landfill Tax.

It was apparent to waterway related organisations that, in conjunction with guidance on disposal, a need also existed for contemporary guidance concerning good practice in inland dredging, embracing all aspects, including initial planning, consideration of environmental matters and implementation of dredging works, whether by contract or direct labour. This report addresses these matters.

1.2 OBJECTIVES

The objectives of the report are:

- to provide an overview of the nature, scale and extent of dredging operations on inland waterways in England, Wales and Northern Ireland, undertaken for navigation, flood defence, nature conservation and amenity purposes
- to categorise dredging, sediment clearance and treatment operations in terms of mechanical techniques and environmental effect and to determine the relative costs associated with each category
- to review and assess current techniques and to describe their advantages and disadvantages under different circumstances

- to assess the environmental impacts and attributes of the techniques in terms of water quality, ecology and protection of the environment
- to propose suitable contractual arrangements and obligations associated with dredging operations and the contractor-client relationship
- to give guidance on good practice, assess the scope for improvement in operational practices and suggest areas for improvement
- to identify and prioritise future research and development needs.

1.3 SCOPE AND APPROACH

This report embraces all significant aspects of inland dredging, other than disposal, which is covered in CIRIA (1996a). Its main emphasis is on maintenance dredging, but the general principles and methods described can, in many cases, also be applied to capital dredging. Where capital dredging requires a radically different approach the alternatives are discussed. The report does not cover the separate control of weed growth, or other vegetation in inland waterways, except by the removal of enriched sediments.

The technical level of the report is aimed primarily at graduates and technicians who, with little knowledge and experience of inland dredging, are seeking guidance. However, the report also addresses some matters at a higher level, which it is hoped will be helpful to experienced professionals.

For the purposes of this report, the term 'inland dredging' is limited to rivers, canals, ponds, broads and small lakes. It does not include estuaries, marinas, or mineral extraction. The task of dredging is addressed under the following headings:

- task evaluation and definition
- dredging techniques
- treatment techniques
- selection of methods
- contractual matters
- health and safety
- environmental effects.

The scale and nature of current practice in England, Wales and Northern Ireland, was assessed by a combination of literature and database searches; interviews and information exchange with selected owners, authorities, operators and contractors; and by a circular questionnaire.

1.4 READERS' GUIDE

This report examines the nature and extent of current practice and provides a guide to good practice.

The results of consultation and a literature review are reported in Chapter 2 to provide an overview of current practice, plant and methods and the main areas of difficulty presently encountered. Those areas where improvement is seen to be desirable are identified. Particular methods or processes are not discussed in detail at this stage, but are described in later chapters. For example, methods of dredging are described in Chapter 5.

Guidance is provided to meet the needs of those involved in dredging inland waters. In order to identify the available and optimum dredging solutions a sequence of assessments and decisions are necessary. The reader is guided through this process in Chapters 3 and 4.

The treatment of some matters, such as the chemistry of treatment processes is, of necessity, superficial. These are complex matters which are developing rapidly. Contemporary information should be obtained by reference to specialists, or recent technical papers and publications.

Chapter 3 provides a general overview and introduction to the subsequent sections, to guide the reader through the report, or to direct him or her to a particular section when a specific task is to be performed in isolation.

Chapters 5 to 9 are intended to be stand alone chapters to which the user should refer for guidance on specific matters.

1.5 COMPLEMENTARY PUBLICATIONS

Specific publications are of special relevance to this work or, in the case of CIRIA Report 157, are by design complementary to this report. For example:

- BS 6349 Part 5: Code of practice for dredging and land reclamation.
- CIRIA (1996a) Report 157 *Guidance on the disposal of dredged material to land*
- PIANC (1990) Supplement to Bulletin No. 70, *Management of dredged material from inland waterways*
- PIANC (1992) *Beneficial uses of dredged material - a practical guide*
- RSPB (1994) *The new rivers and wildlife handbook*
- Anglian Water Authority (1992) *Conservation guidelines for river engineers*
- Nature Conservancy Council (1983) *Nature conservation and river engineering*
- Bray *et al.* (1996) *Dredging: a handbook for engineers.*

2 Current practice

2.1 INTRODUCTION

A general picture of current practice has been obtained by a combination of literature review, interviews and the circulation of questionnaires. All major organisations involved with the dredging of inland waterways in England, Wales and Northern Ireland have been consulted. In addition, interviews have been conducted in France, Belgium and Holland with the intention of identifying any significant differences in practice relative to the UK.

The consultation process has produced a wealth of information which is too extensive to report in detail, but a summary is provided in Tables 2.1, 2.2, 2.3 and 2.4.

Significant inconsistency has emerged in the information gained by interviews and from questionnaires. This is believed to be due in part to interviewees and consultees not attaching a high level of importance to accuracy when providing figures. However, so great are the discrepancies in many instances that it must be concluded that some organisations, or divisions of organisations, have only a very limited understanding of the extent and cost of their inland dredging operations. For this reason the statistics reported in this report should be treated with caution.

There appears to be a need for improvement in the methods of monitoring, recording and analysing the need for and results of dredging (see the recommendations outlined in Chapter 10, Section 10.1).

2.2 OVERVIEW OF THE NATURE AND SCALE OF OPERATIONS IN ENGLAND, IRELAND AND WALES [2]

The relative scale of inland dredging in terms of volume and value is headed by river and watercourse dredging, followed by canal, lake and pond dredging and finally, reservoir dredging..In England and Wales, rivers and watercourses are generally the responsibility of the Environment Agency, or Internal Drainage Boards. In Northern Ireland they come under DANI. Canals and navigable waters are predominately the responsibility of British Waterways, but the Environment Agency and DANI also maintain substantial lengths. The Manchester Ship Canal Company is responsible for the 58 km length of the ship canal. The Broads Authority is responsible for 200 km of navigable inland waterways. The ownership of ponds, lakes and reservoirs is diverse.

In England, Wales and Northern Ireland, a total of some 60,000 km of inland waterways are maintained by the various bodies referred to above. Not all of these require dredging and only few require very regular dredging, but approximately 15,000 km do require maintenance dredging at varying intervals. These give rise to an annual dredging production in the range 7 to 10 million tonnes of sediment at a cost of between £15 and £20 million.

---

[2] In Scotland, rivers and watercourses are the responsibility of the Scottish Environmental Protection Agency.

In recent years capital dredging has been uncommon. The results of consultation indicate that it represents only 10% by value and 8% by volume of the total dredging activity on inland waterways in England, Wales and Northern Ireland.

It is the disposal of the dredged material which is seen to be the more technically challenging and expensive task (CIRIA, 1996a). Recently, disposal problems have been magnified by increasing environmental and legislative constraints. This trend continues. Where dredging is for the specific purpose of environmental improvement, as in the case of the removal of sediments contaminated by past industrial activity, special treatment and disposal methods must be employed and costs may be particularly high.

Dredging costs recently experienced by British Waterways have ranged from £3 per cubic metre for simple dredging and disposal to £50 per cubic metre for difficult work involving the dredging and treatment of contaminated sediments.

Table 2.1 *Summary of statistics collected on the nature and scale of dredging*

	Capital		Maintenance	
	Contract	Direct Labour	Contract	Direct Labour
Value (£)	2,200,000	360,000	9,500,000	8,600,000
Volume (m3)	260,000	15,000	1,000,000	2,300,000
Length (km)	14	1	6,200	9,000

2.3 OVERVIEW OF CURRENT PLANNING OF WORK

With notable exceptions, such as the Manchester Ship Canal, the strategic planning of maintenance dredging for canals and rivers has often been haphazard. There is evidence that the situation is improving, but improvement is patchy and further progress is necessary to identify and achieve optimum arrangements. (see Section 3.4).

Planning is commonly hampered by a lack of detailed information concerning rates of siltation, in terms of sediment volume, or loss of depth. This may improve with the availability of new survey methods which allow rapid data gathering and interpretation, particularly in navigable waterways, (see Section 7.3.2). However, in some areas a reluctance to adopt new methods is apparent.

Dredging of rivers and canals is driven primarily by need. Work is carried out when it is recognised that local loss of channel section is approaching the point when the risk of flooding, or restriction to navigation, is unacceptable. The Broads Authority has a target of dredging navigable waters every 10 years and British Waterways have navigable depth targets for all of their waterways, but these targets are not consistently achieved due to a lack of finance.

On the other hand, the Broads Authority has for 15 years been implementing the dredging of shallow lakes as part of a lake restoration and research strategy, resulting in clear water conditions and the reappearance of many rare and protected aquatic species.

Whenever possible, dredging is planned to minimise adverse impact on the local environment, for example, by avoiding the nesting season for waterfowl, or the growing season for arable crops. However, other factors may conflict. Dredging may not be practicable if, for example, river level or flow is high.

The planning of dredging in canals and rivers usually includes consultation with a range of environmental groups, including English Nature, Royal Society for the Protection of Birds (RSPB), local fishing and environmental groups (see Chapter 9).

British Waterways have an environmental unit (Environmental & Scientific Services) which is located at their Gloucester office. This provides guidance to planning and operational staff when requested, particularly if work is planned in an area of recognised sensitivity, such as an SSSI. They have also developed an 'Environmental Code of Practice' for internal use.

Similarly, the Environment Agency have conservation guidelines for internal use provided by *Conservation guidelines for river engineers* (Anglian Water Authority, 1982).

2.4 GENERAL APPLICATION OF PRIMARY LAND-BASED EXCAVATION EQUIPMENT

A wide variety of dredging equipment is available for use on inland waterways. Machines that work from the bank and usually travel on tracks include hydraulic backhoes, draglines and grabs. Floating equipment includes hydraulic backhoes and grabs, rope-operated grabs, bucket dredgers, suction dredgers, ploughs and water injection devices. Each dredging method is described in Chapter 5. Significant regional variation occurs, the types of equipment in use being dictated by local need. As a result, some of the equipment described in Chapter 5 may not be readily available in all areas.

Analysis of interviews and questionnaires indicates that approximately 90% of the plant used for dredging inland waterways is land based; tracked backhoes (see Section 5.3) are the most common, with a share of over 70% of land-based dredging.

Land-based dredging plant is dominant where the areas to be dredged can be reached from the land, where bank access is possible and dredged arisings can be deposited on to adjacent land, or loaded into trucks or dumpers. If some or all of these conditions are satisfied, it is usually the case that land-based dredging will be the least expensive method. This general rule may be affected if the dredged arisings must be disposed of at a location more easily reached by water. Transport by floating hoppers may then be less expensive than transport overland.

Internal drainage authorities, which are mainly concerned with the creation, or maintenance, of relatively small drainage ditches and channels, employ land-based excavators exclusively.

The Environment Agency, which is responsible for watercourses and rivers with a wider range of dimensions, of which only a small proportion are navigable, employs predominately land-based excavators. Floating dredgers are used in a relatively small number of locations where special conditions prevail.

In contrast, British Waterways and the Broads Authority, whose waters by definition are usually navigable, employ floating dredgers more extensively.

There is a simple relationship between the optimum type and size of dredging plant employed and the dimensions of the area to dredged. Machine outreach from the bank is usually limited to the maximum distance over which the free-standing machine can cast and recover the digging bucket. The maximum outreach is related to the weight of excavator. Those in common use for inland dredging have a maximum outreach of about 20 m (see Chapter 3). A rare exception is found in systems which rely on pulling the bucket with one or more winches between points on opposite banks. Such systems may operate over widths in excess of 100 m (see Section 5.7).

A further exception is the use of tracked machines which are designed to work standing in water within the area to be dredged. By this method the maximum width which can be dredged may be almost doubled.

Where land-based dredging plant is used, the most common types are track-mounted hydraulic backhoes and draglines. Earlier, draglines were the most commonly employed method for river dredging. Improvements in the design and construction of hydraulic backhoes has resulted in a progressive improvement in performance, particularly in respect of out-reach. This has reduced the role of the dragline, but it remains a useful tool for some applications, particularly those requiring long reach.

A survey of the current market shows that the dredging plant which is owned and operated by the various inland drainage and navigation authorities consulted is predominately land based. Percentages by type are given in Table 2.2. However, these figures should be treated with caution because of distorting factors. For example, whereas most dredging for the Broads Authority and British Waterways is by floating plant, much of the work is carried out by Contractors using their own plant.

Table 2.2 *Land-based plant - percentages of total number of dredging plant employed*

All land based plant	83%
Hydraulic backhoe*	67%
Hydraulic backhoe, combined rope and hydraulics (VC's)	3%
Dragline	14%

* Only a small percentage of backhoes are standard machines

2.5 GENERAL APPLICATION OF PRIMARY FLOATING DREDGING EQUIPMENT

The use of floating dredging plant is commonly employed in canal, broad and lake maintenance. In the dredging of lakes and broads the bank to bank width is usually too great to allow all areas of the bed to be reached from the banks. Ponds, which by definition are less than one acre in area, may be dredged by land-based or floating dredgers, depending on the maximum dimensions and local circumstances.

The more common use of floating plant for the dredging of canals is due primarily to restricted bank access, which often is either too narrow for access by appropriate land-based plant, or does not provide adequate space for bank-side disposal. There may also be a natural pre-disposition for navigation authorities to employ floating plant which can operate freely within the areas of its ownership and jurisdiction.

No particular type of floating plant dominates the dredging of inland waterways. Grabs, backhoes, bucket dredgers and suction dredgers are all in fairly common use (see Table 2.3).

The excavating characteristics of land-based and floating plant may also be found in special amphibious plant. Many different designs are available. Those in most common use comprise a main pontoon structure which houses the power unit fitted with a means of excavation. Excavation may be by hydraulic backhoe, grab, or by pumping. Excavation by backhoe attachment is most common in the UK, whilst pumping methods are more common in continental Europe.

Travel of amphibious plant may be by means of powered tracks fitted at the base of vertical spud-like legs, or, more commonly in UK inland dredging, the main pontoon may be fitted with hydraulically powered arms, or legs, which terminate in large buoyancy pods. Such machines are able to work in water or in bog conditions. A particular advantage is their ability to climb into and out of the water under their own power. This may simplify access to and from the working area, or diversion around restricting structures (see Figure 2.1).

Table 2.3 *Floating plant - percentages of total number of dredging plant employed*

All floating plant	17%
Hydraulic backhoe*	4%
Grab	5%
Bucket	4%
Cutter suction	2%

* Temporarily or permanently mounted on pontoon, or barge

2.6 TREATMENT METHODS

Treatment may be necessary, or desirable, for a variety of reasons. De-watering is necessary to reduce bulk, or render the material suitable for transport, or agricultural use. It may also render the material suitable for acceptance for disposal at a licensed site which is closer to the area of dredging than if disposal is restricted to a site that accepts wet waste. At one extreme, de-watering may be achieved simply by drainage following deposition on bank-side land. At the other extreme, the method of de-watering may be sophisticated and expensive, such as by belt-press (see Chapter 6).

Figure 2.1 *Amphibious dredger working in bog conditions*

De-watering is unlikely to reduce the materials level of contamination and may in fact result in the concentration of contaminants to such a level that special disposal methods are necessary.

As a general rule, treatment, other than basic de-watering and screening to remove debris, is not necessary for dredged material arising from land drainage ditches, channels and rivers, lakes, or broads. Exceptions may arise downstream of current or past industrial activity, where the bed material may be contaminated.

In operations which only involve clean dredged material, screening to remove debris also may not be essential, with reliance being placed instead on the hand-picking of large or potentially harmful items by operatives. It is not common for material routinely dredged from rivers and watercourses to be screened. However, when dredging canals, screening is more common and is becoming increasingly so. Why the practice for rivers differs is not completely clear, but appears to arise for a variety of reasons, including differing statutory rights concerning access and disposal. Recently, major canal clearing projects have been concentrated in urban areas, where pollution by debris is most common. The increased use of agricultural land for material disposal also dictates that debris be removed.

Debris, or other oversize material which is separated by screening, must be disposed of separately.

Other forms of treatment are necessary if the dredged material is significantly contaminated by harmful substances. The threshold level of contaminant concentration above which treatment is necessary is dependent on the particular contaminate, or contaminants. It has also been the case that threshold levels varied between different local Waste Regulatory Authorities. The Environment Agency are now (June 1996) working to produce national standards for waste classification, but in the interim regional inconsistency may continue. Limits may also be influenced by the proposed method and place of disposal or end use. Treatment is not yet widespread and no statistics have been collected concerning the use of different methods.

2.7 DISPOSAL AND TRANSPORT OF DREDGED MATERIAL

2.7.1 Disposal

The disposal of dredged material has been strongly affected by the introduction in 1994 of the Waste Management Licensing Regulations. This has had the effect of promoting certain disposal options over others, restricting the volume of material disposed of at certain sites and, in the case of contaminated material, of increasing costs. Disposal options in common use are:

- bank-side spreading
- spreading on agricultural land, with benefit to agriculture
- beneficial re-use or use for ecological improvement, e.g. land reclamation
- hauling or pumping to own dedicated tip or lagoon
- hauling to commercial licensed waste disposal site.

Material arising from the dredging of drainage ditches, channels and rivers, is most commonly disposed of by spreading on the adjacent bank, or on adjoining agricultural land. Where applicable, dredged material may also be used to improve flood embankments. The latter method is sometimes used by the Broads Authority, working in co-operation with the Environment Agency, to their mutual advantage. Subject to certain conditions (see CIRIA, 1996a), these methods are granted exemption from licensing under the Waste Management Licensing Regulations. As a consequence, it is probable that reliance will be placed on these methods wherever appropriate.

The volume of material arising from the dredging of individual lakes and broads often is too great for simple bank-side disposal. If so, then the choice is usually between spreading on agricultural land, for which exemption is necessary, or disposal off-site, for which a licence may be required. Subject to the material being clean, other possibilities may be found in beneficial uses, such as composting and conversion to top-soil.

When dredging canals, or other navigable waterways, bank-side disposal is less common and removal off-site to a designated disposal area is often necessary. This may result in cost increases in excess of 100% relative to bank-side disposal. The disposal methods employed by British Waterways, as reported in interviews and questionnaires returned by three British Waterways regions, break down as follows:

• exempt disposal to bank-side or agricultural land	72%
• own dedicated tip, or lagoon	10%
• off-site to third party tip	18%
• other	0%

These figures are very approximate averages for 1995 only. The percentages may change from year to year and marked variations are reported between regions. Bank-side disposal predominates in all three regions, but disposal at a dedicated owner's tip ranges from 5% to 25%. Disposal at an off-site third party tip ranges from 5% to 45%.

For rivers and watercourses, comparable figures to those provided above for canals, as reported by the NRA (now the Environment Agency), were as follows:

• exempt disposal to bank-side or agricultural land	71%
• own dedicated tip, or lagoon	4%
• off-site to third party tip	21%
• other	4%

The above figures for rivers and watercourses are averages derived from the analysis of questionnaires returned by 19 regions, or sub-regions. Individual regions exhibit substantial variation which is mainly influenced by the character of the area and methods of dredging employed. For example, in Cornwall, where land-based methods of dredging are employed exclusively, 99% of disposal is bank-side, whereas in the Thames Region, Tidal South East, which covers the river Thames downstream of Kingston, 100% of disposal is off-site. In general, it appears that in freshwater areas and steep catchments, bank-side disposal is normal, whereas in lower reaches and tidal areas, disposal to dedicated lagoons, or off-site, is common.

Where material must be disposed of at licensed landfill sites, apart from substantial added cost, problems may be caused by the relatively low daily tonnage rates which many tips are licensed to accept and the small number of sites licensed, or willing, to accept wet waste. - wet being defined as with moisture content in excess of 40%.

2.7.2 Transport

With the exception of bank-side disposal, dredging by land-based plant is most commonly accompanied by over-land transport in trucks or dumpers, these being better matched to the rate of production of land-based plant, which is usually higher than that of floating plant.

When dredging is by floating plant, transport by hopper barge is more common because land-based transport is not easily loaded by floating plant. Occasionally transport is by pumping through pipelines.

As seen in the statistics relating to disposal, most material is deposited on banks. Where transport is necessary, the method used, as reported in interviews and questionnaires, are shown in Table 2.4.

Table 2.4 *Analysis of transport methods used for the disposal of dredged material*

	All parties	NRA	BW
by pipeline	12%	5%	20%
by hopper barge	16%	11%	45%
by dump truck	13%	16%	0%
by road truck	50%	62%	35%

2.8 PRINCIPAL PROBLEM AREAS

The results of interviews with inland waterway authorities and specialist dredging contractors, and the analysis of questionnaires, has highlighted a range of problems. These vary to a limited degree between drainage and navigational waterways, but the more important problems are common to both. In general, no particularly severe problems have been identified, other than the very recent changes in respect of disposal resulting from the Waste Management Licensing Regulations (see CIRIA, 1996a).

The consultation questionnaire divided problems into two areas, operational and administrative. Those problems which were mentioned most commonly, either in interviews or on questionnaires, are outlined in the following sections.

Operational problem areas

Operational problems are listed below in descending order of difficulty, or frequency encountered, as reported by industry; i.e. access for dredging and transport plant was most commonly quoted as a source of difficulty:

- access
- disposal
- site security
- water quality
- mechanical failure
- safety, plant limitations, equal ranking
- supervision
- labour
- water supply.

Administrative problem areas

Administrative problems are listed below in descending order of difficulty, or frequency encountered, as reported by industry:

- environmental
- legal, fiscal - equal ranking
- health and safety
- contractual.

The level of difficulty associated with the operation and administrative matters listed varies considerably. Those problems which are persistent, or particularly significant, are described below.

Access

Access restrictions are commonly caused by weak ground, physical obstruction, growing crops or grazing cattle, or relatively rarely, by legal constraint when the land is privately owned and no statutory right of way is vested in the authority wishing to gain access.

It is usually the case that those organisations which have a statutory duty to maintain waterways also have a right of entry to land for the purposes of carrying out the work of maintenance. Subject to giving an appropriate period of notice of the intention to carry out work, the authority may be protected against claims for unavoidable damage, such as crop damage. However, it may nevertheless be prudent, when possible, to programme work over agricultural lands at times that avoid or minimise damage.

Access to the banks of minor waterways is often impeded by encroaching development (see Figure 2.2). This may render the use of simple land-based methods impossible. The level of control of bank-side development may vary depending on the classification of watercourses, or detail of land acquisition in the case of canals. Even where bank-side development is the subject of control, unauthorised development may occur, the

Figure 2.2 *Limited bank access*

removal of which may be legally complicated. Such problems are most common in urban and industrial areas.

Some authorities report that accidents involving bank-mounted excavators are a source of concern. These usually involve either machines becoming bogged down when traversing soft ground, or slipping into water or cuts as a result of bank edge failure, or lack of care by the machine operator. (see Figure 2.3).

The security of access should be carefully assessed during the initial inspection of the site (see Section 4.1). Risk to life and property does not end with the initial failure, but continues through the period of machine recovery, during which substantial and unpredictable forces may be generated when lifting or pulling. It is important that machine recovery operations are carefully planned and supervised (see Section 8.4).

Contamination

When bed materials are contaminated by any substance which is potentially harmful to human health, or to the natural environment, the processes of dredging, transport and disposal are subject to important restrictions. Generally, the severity of restriction will increase with the severity of contamination. Restrictions may be imposed through a variety of legislation (see Section 9.2).

Material which contains a significant level of contamination may only be disposed of in an approved manner at an appropriate licensed disposal site. Treatment, though usually expensive, may remove or reduce contamination to a level which permits a range of disposal options and reduces the overall cost of disposal.

The dredging process is less regulated than disposal. However, methods that result in the release of a significant level of contaminants to the local aquatic environment will adversely affect water quality, which is the subject of strict control.

Figure 2.3   *Accident resulting from bank edge failure*

If transported, contaminated material must be adequately contained such that spillage or wind-borne dispersal is prevented. For road transport, this usually will dictate the use of special vehicles.

Debris

Debris is a common problem in canals or watercourses which pass through urban or industrial areas. In such areas some level of pollution by debris is to be expected. Debris may also arise from natural sources, the most common being timber, in the form of fallen branches or trunks.

The effect of debris is to restrict the range of dredging methods which may be economically employed, unless the debris is previously removed, and to complicate disposal. For example, dredged material which contains substantial quantities of debris, natural or otherwise, usually cannot be economically dredged by suction methods.

To avoid the risk of harm to farm livestock or machinery, debris should be removed from dredged material when disposal is by spreading on agricultural land. Similarly, if contaminated dredged material is to be treated it will usually be necessary to remove any debris first.

Disposal

The main problem associated with the disposal of clean dredged material is simply that of finding land which can be used for the purpose.

Unless land used for disposal is bank-side land, or agricultural land on which dredged material may be spread without adverse effect, or the end use qualifies for exemption under the Waste Management Licensing Regulations, the disposal site must be licensed and will be subject to the Regulations. Compliance with the Regulations will usually result in increased costs.

Such sites are few in number because of the high level of regulation attached to a licensed disposal site. This scarcity, which is likely to persist, will generally result in greater transport distances and hence higher costs.

Financial

Most agencies and authorities that undertake the maintenance of inland waterways operate on the basis of pre-determined annual budgets. This may result in the inefficient planning and execution of work. A prime example is that of substantial work which will be most efficiently and economically executed in one continuous operation, but which must be divided so that the annual budget is not exceeded. Separate contracts or operations result in an increase in overall cost, which may be substantial.

2.9 CONTINENTAL EXPERIENCE

Visits and interviews were conducted to determine inland dredging practice in Belgium, France and Holland. In general, the methods of dredging employed were found to be broadly similar to those in the UK, though influenced by the larger dimensions of canals and rivers. The use of floating dredging plant is more common than in the UK, with backhoe, bucket, cutter suction and grab dredgers all being in common use.

In some technical and contractual matters, quite significant differences in practice were found.

France

Dredging of inland waterways is almost exclusively by specialist Contractor, with an estimated 95% of work being by contract.

Dredging contracts are usually measured contracts with measurement being based on the volume removed, but with some limit on the amount of paid over-dredging. It is generally the case that the volume of over-dredging included in the measurement shall not exceed 2% of the theoretical volume within the design profile. However, regional variations in contract detail occur due to separate administrative regions and widely varying site conditions.

Contracts are often term contracts, typically of 3 years duration, during which period the Contractor is charged with maintaining a specified minimum depth, or profile.

Arrangements for the disposal of dredged material may be specified, or left to the Contractor, the latter being most common. There is general compliance with the European Waste Directive. Finding suitable sites for disposal and obtaining consent is seen as an important and growing problem. Locally, it is necessary to persuade the Mayor to permit disposal, but lobbying by environmental pressure groups often results in failure. Wastewater draining from disposal sites is sampled and tested by state officials.

The beneficial use of dredged material is not common in maintenance works, but is encouraged in capital works, where the material arising may be better suited to beneficial use.

Routine treatment is not common. Screening is not usually necessary, but hydrocyclones may be used to separate sand where commercially viable.

Dredging costs are highly variable across the country. For the Seine, typical costs for simple dredging and local disposal are £13/m3. This rises to about £40/m^3 if treated (subject to type of treatment), and further to £65/m^3 for disposal at a licensed site.

Belgium

All dredging is by contract. Preference is given to grab and bucket dredgers because of the problem of added water content if suction methods are used. Specialised dredging plant is rarely employed in the canal system due to the fact that water quality generally is so low that special methods designed to minimise the release and spread of sediments are unjustified.

Small canals, of less than 5 m width, are usually dredged by tracked, or wheeled, excavators working from the bank. A 5 m wide bank width is preserved by law for access and disposal, unless material is contaminated, when it must be disposed of off site. The employer specifies and provides a land area for the disposal site.

Work is measured and valued by volume by pre-dredge and post-dredge survey. Maintenance of 550 km of canal system requires the dredging of approximately 800,000 m^3 per year. This volume reduces by between 30% and 50% due to drainage and consolidation in disposal areas.

The interval between maintenance campaigns is increased to minimise local disturbance, increase productivity and reduce costs. To date, the main purpose of maintenance dredging has been to facilitate navigation or water capacity. However, environmental 'clean-up' dredging is of growing importance.

The main thrust of management effort is directed towards improved understanding of siltation and pollution, improved monitoring and forward planning.

Dredged material is no longer perceived as waste, but as a potential resource which, whenever possible, should be used beneficially.

The past perception that heavy metals are one of the most important sources of contamination has changed, with agricultural chemicals, oil and sewage, now being targeted as more important pollutants. It has been found by large-scale testing and monitoring that heavy metals generally remain bonded to the dredged soils and do not leach from disposal sites.

The *in situ* treatment of organically contaminated sediments has been carried out on a trial basis. Treatment has involved the injection of the bottom sediments with a cocktail of oxygen fixates, micro-organisms and nutrients. To date the results are inconclusive. However, when similar methods are applied to dredged materials placed ashore, improved control is possible and the method is reported to have been 99% successful.

A variation on the method, whereby the soil dry matter is diluted by 50% and injected with oxygen and nutrients to encourage biological activity, has also been found to be quite successful, subject to low heavy metal content.

A success rate of about 40% in removal of free heavy metals has been achieved by the addition of chemicals during pumping to a lagoon. This causes release of heavy metals to the water which is then treated separately.

Dedicated commercial plant for the treatment of contaminated soils are also in operation with others planned. These are not exclusively for dredged material. Typical maximum throughput is of the order of 30 tonnes per day. Capital investment in the plant amount to some £10 million. The multi-stage process involves dilution of the soil mixture with water and hence the usual incentive to minimise water content when dredging does not apply unless by so doing transport to the plant is simplified.

A novel beneficial use which, although still under trial, has achieved some success and acceptance, is the mixing of granular dredged sediments with pig manure in ratios ranging from 2:1 to 1:1 for use in agriculture for soil improvement. Good and improved crop yields have resulted.

Holland

Approximately 1,500 km of canals are maintained, all by contractors. Large canals are the responsibility of a central state organisation, Rijkswaterstaat. Small canals are the responsibility of the local municipality, or water boards. Approximately 20 million m^3 are dredged in maintenance, of which some 9 million are contaminated. As a consequence research and development is directed primarily to improving methods of dredging contaminated soils.

Special equipment has been developed for the dredging of contaminated soils with a vertical accuracy of plus or minus 50 mm. These are mainly suction dredgers with special shielded cutterheads and suction intakes, but no particular type of dredger has yet emerged as the favoured method for dredging contaminated soils. Pumped concentration of 80% *in situ* material is claimed using this equipment. Transport water is used in the treatment process and re-circulated. Local increase in turbidity due to dredging is minimal.

Following hesitant acceptance, water injection methods (see Section 5.4) are gaining favour in appropriate situations. The planned method is to lower bed levels by dredging to create a receiving area, to move contaminated sediments into the deepened area by water injection and to cap it with clean material. Subject to good control and optimum injection rate, unwanted dispersion of mobilised sediment has been found to be within acceptable limits.

National standards are established for categorising contaminated sediments and determining dredging, treatment and disposal requirements. The process of dredging requires a licence. The re-circulation of water used in the dredging process is strictly controlled.

A large-scale separation (sand from fines) treatment plant is under construction at the 'Slufter' disposal site located at the Hook of Holland. This has a design capacity of 80 tonnes per hour.

Areas for use in the disposal of dredged materials are subject to planning permission.

Inland dredging is most commonly measured by volume by pre-dredge and post-dredge survey. Contracts are awarded by a two-stage process. Stage 1 is a technical appraisal and stage 2 financial. Only those tenders deemed to be technically sound, following appraisal at stage 1, are subject to financial evaluation.

Technical areas found to pose particular problems, are those of identification, at the planning stage of work, of the levels of pollution by debris and of contamination. Regardless of the dredging method, the presence of debris in significant quantity greatly detracts from the ability to dredge contaminated sediments without significant release and dispersion.

A survey system for identifying debris has been tested, but found lacking. Development continues.

Systems are under development for the spot measurement of contamination. These show greater promise. A system able to measure the level of a few contaminants has been successfully tested. This recovers a sample and completes limited analysis on site within 5 minutes.

Development of a suction dredger, preferably fitted with de-gassing equipment insensitive to debris, is seen as desirable.

3 Project evaluation and implementation

3.1 INTRODUCTION

This chapter provides an overview of the process of project evaluation and implementation and directs readers to those sections where relevant issues are covered in more detail.

The optimum solution, or alternative solutions, to a specific dredging problem can only be identified given a thorough understanding of the scale and nature of the required task and the physical, environmental and regulatory constraints which will, or may, exert influence. Only when the task has been adequately defined can the optimum solution be identified.

The basic steps leading from initial recognition of a problem through to implementation of a dredging solution are illustrated by Figure 3.1. The problem may arise routinely, or unexpectedly (see Chapter 4), but in either case the solution to be achieved should be defined in terms of specific objectives. For example, in the case of a navigation canal, the primary objective is likely to be the restoration of a specific water section in terms of water depth and width.

Initial consideration of a particular problem should include the question 'Is dredging the only solution?' If not, is it the optimum solution?

If it is decided that dredging is the required solution, or at least one which should be considered in parallel with alternatives, the next stage of evaluation is complex. Figure 3.1 illustrates six basic matters to be considered in order to progress to an optimum dredging solution. These are interactive and should each be considered in relation to the others in an iterative manner.

The six basic matters are presented in a circle because there is no particular starting point in the evaluation which is consistent for all types of problem. Sometimes the method and place of disposal of the dredged material may be the most important single factor and hence should be considered first. For other sites, it may be that access restrictions, or the character of the materials to be dredged, dictate that a particular method of dredging be employed.

It is essential that an iterative process of evaluation be adopted and it is recommended that an initial rapid assessment be made using Chart 5 (Figure 3.8), with the object of identifying the main constraints or influences on any solution.

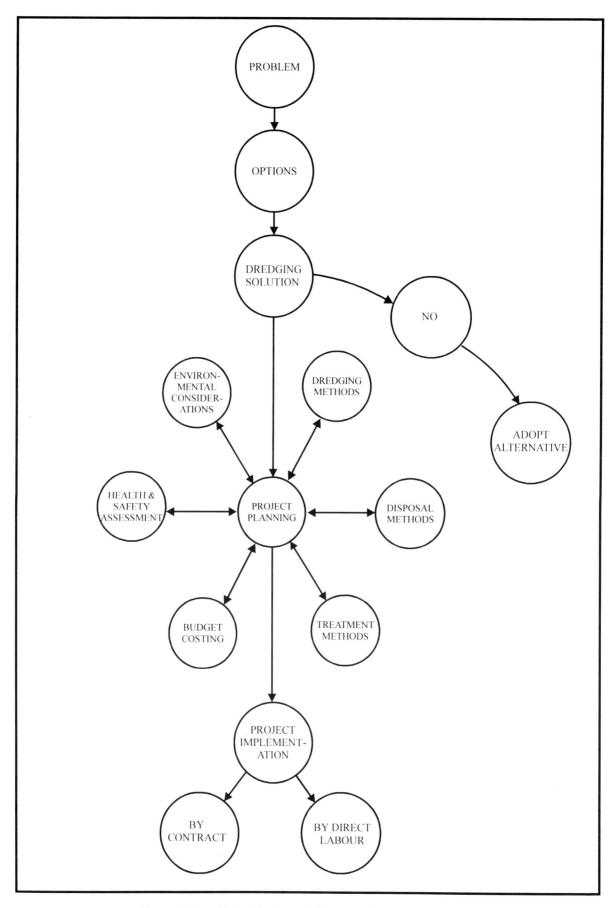

Figure 3.1 *Project implementation procedure*

3.2 OBJECTIVE - REQUIRED LEVEL OF SERVICE

In most cases, the level of service to be provided by an inland waterway will be defined in terms of a set of dimensions. The dimensions of a drainage channel are based on the need to pass a maximum flood flow with a specified return period, perhaps 1 in 10 years. For canals, the dimensions may be based on providing a specified minimum under-keel clearance and passing width which are appropriate for 90% of traffic.

Thus the object of dredging will vary from site to site:

- for canals, broads and navigable rivers, the primary objective usually will be to permit safe and unimpeded navigation, but where a canal serves to transport water, or land drainage, dredging may be necessary to accommodate a specified flow
- for rivers, the object will usually be to accommodate a particular maximum flow, or to concentrate low flows
- for ponds, lakes and broads, an increase in water depth, capacity and water quality may be the primary objectives to achieve or maintain an environment of a particular character
- for reservoirs, the object will be to restore lost capacity, or to clear a passage to sluices or intakes.

The level of service to be provided should be reviewed during the planning phase. It should be clear that in a canal it will be uneconomic to provide for an unusually large vessel if navigation by that size of vessel is rare. Similarly, the maintenance of a river section to pass an extreme flood flow is unlikely to be justified if flooding will only occasionally inundate meadow land. These examples are unlikely to be considered in the prevailing stringent economic conditions, but the importance of regular re-assessment of the level of service which can be justified remains.

For all projects, but particularly ponds, lakes and broads, an additional and often primary objective of environmental improvement may exist, either by the removal of contaminated sediments or debris, or by altering the physical characteristics of the aquatic and marginal habitat. Usually it will be necessary to consider environmental factors in parallel with the required dimensions.

The needs of canals, rivers, ponds, lakes and broads may not be the same, but in determining the appropriate level of service, some factors are likely to be common, particularly where dredging is for maintenance purposes.

3.3 ALTERNATIVES TO DREDGING

Solutions other than dredging may exist. The most satisfactory of these may be to do nothing. This may appear nonsensical, but in fact examples of unnecessary dredging are not uncommon.

The reasons for unnecessary dredging may be obscure, but may stem from dredging plant and operatives being available and the perception that they should be gainfully employed. Perhaps the dredging plant was originally acquired for capital works, or for a maintenance dredging task which was historically more extensive. If doubt exists concerning the need for dredging, it is advisable to cease dredging for an appropriate period, perhaps several years, and in the interim, to monitor bed levels and cross-sections by regular survey.

When the process of dredging and disposal can only be satisfactorily achieved if accompanied by treatment, other than the most basic, such as screening, it will be prudent to question the wisdom of dredging and to consider carefully alternative solutions.

The alternative of doing nothing will not only save money, but may in the long term have less impact on the environment. For example, in rivers and estuarial waters which were once heavily polluted, but which now are relatively clean, uncontaminated sediments may be deposited over polluted sediments, effectively capping and isolating the contaminants. Dredging will expose and may release and disperse the contaminating elements. Hence, if the primary object of dredging is to improve the environment, rather than to improve water depth, to do nothing may in some instances be more beneficial.

In most cases, the need for improvement of waterway depth or section may be unavoidable, but a solution other than dredging may nevertheless exist. For canals, raising water level, if feasible, may provide a short-term solution. Increasing flow rates, again if feasible and economic, may eliminate, or reduce, the rate of siltation. Reducing sediment input, perhaps by routing feed water via a settling lagoon, or by feed over a raised weir level, could reduce or even eliminate the need for dredging within some canals, ponds and lakes.

Where shoaling is at isolated locations within an otherwise stable regime, re-suspension and dispersal of the shoal sediments may provide a low cost solution. This may be done in a variety of ways (see Section 5.4).

There may be other alternatives to dredging, but the usefulness of these are site dependent and can only be assessed on a site specific basis following definition of the task (see Chapter 4).

3.4 PROJECT PLANNING

3.4.1 Introduction

Maintenance dredging is necessary when water depth or the waterway cross-section is no longer sufficient to provide the required level of service. In most cases rate of change from a satisfactory to an unsatisfactory level of service will be gradual, though the rate may vary with season or climatic conditions. It follows that to foresee the need for maintenance and to properly plan an appropriate programme of work, periodic monitoring of conditions is essential.

The effectiveness of forward planning decisions is highly dependent on the quality of information available. Essential information includes the more or less fixed parameters of maximum and minimum bed levels, the rate of sediment input and deposition, which may vary, and the rate of loss of depth.

The variable factors, loss of depth and section, can only be determined by means of regular survey and analysis. Regular need not mean frequent, the optimum frequency being dependent on the rate of change.

Given regular survey data, which should be collected most frequently at critical locations, a simple graphical record which records loss of depth, or section, will serve to illustrate trends in siltation rates and to provide a forecast of the range of dates when action will become necessary (see Figure 3.2).

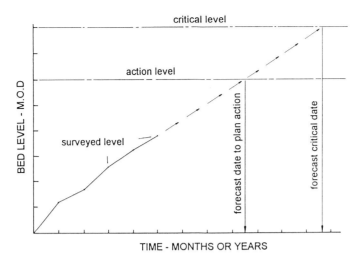

Figure 3.2 *Prediction of dredging need*

In parallel with the need to maintain or restore a particular level of service, the planning of dredging work should take account of the potential impact on the environment, efficiency of the operation and economics.

3.4.2 Frequency and extent of dredging

The optimum frequency with which maintenance dredging should be carried out is dependent on local site conditions and hence each site should be separately assessed to determine the most appropriate interval between dredging campaigns. The objective should be to achieve the optimum balance between functional requirements, environmental protection and economics.

For some waterways, both the cost and environmental impact of maintenance dredging will be reduced if work is carried out less frequently. Maximising the period between campaigns may result in lower unit costs for dredging and the frequency of disturbance of the habitat is reduced. The longer intervening time which is available between campaigns provides greater opportunity for maximum recovery of the ecosystem. However, there may also be an environmental disbenefit in reducing the frequency of dredging (see Box 3.1 and Section 9.3.3), e.g. in the short-term the larger scale operation required for less frequent dredging may be more disruptive than the smaller scale of operation required for the frequent dredging of less material more quickly.

Maximising the interval between campaigns will maximise the volume of dredging to be done. Usually larger quantities will result in lower unit costs for dredging, treatment and disposal. Furthermore, the ratio of plant mobilisation and demobilisation fixed costs relative to unit costs will be reduced when the interval between campaigns is increased. Conversely, increased quantities might require double handling or result in loss of the opportunity to employ low cost disposal methods, such as bank-side spreading, due to exceedance of the limits applicable to an exemption (see Section 4.6). Costs may then be higher.

Box 3.1   *Relevant merits of maximum and minimum intervals between dredging*

Advantages of maximising interval between dredging:

- Period free of environmental disturbance is maximised

- Depth of sediment to be dredged is maximised

- Volume of sediment to be dredged is maximised

- Density of sediment is maximised (minimum water content)

- Ratio of unwanted over-dredging is minimised

- Unit cost of mobilising and demobilising plant is minimised

- Unit cost of dredging is minimised

- Hindrance of waterway users is minimised

Advantages of minimising interval between dredging:

- Scale of environmental disturbance is reduced

- Because the scale of disturbance is less, environmental recovery is more rapid

- Duration of dredging activity is reduced

- Interim loss of water depth, or section, is minimised

- Volume of material to be disposed per unit length is reduced

---

In terms of the preferred extent of dredging Section 9.3.3 **(cross section)** discusses the environmental benefits of dredging short sections during each dredging cycle and of preserving or creating a shallow reed margin to minimise erosion and enhance the habitat.

3.4.3   Canals and navigable waterways

Maintenance dredging in navigable waters is necessary when bed levels are such that the under-keel clearance for the deepest draught traffic is approaching the minimum which is compatible with safe, efficient and unrestricted navigation. Based on known vessel characteristics, this is a fixed bed level for a given minimum water level. If the passage of deep draught vessels is rare, the economics of providing unrestricted access should be assessed.

A further factor of relevance is that the reduced blockage achieved as a result of dredging in turn results in a significant reduction in bow wave wash and consequent bank erosion. It also beneficially reduces the energy necessary to propel a vessel at a given speed.

The maximum water depth, or lowest acceptable bed level, may also be fixed. If the canal is clay lined, over-dredging may damage the lining with consequent water loss. In most cases, severe over-dredging will cause instability in banks or structures and hence should be avoided.

It is important when planning dredging work to give careful thought to the optimum dimensions which dredging is to achieve. Generally, it should not be difficult to determine the maximum and minimum acceptable bed levels. The difference in these levels, when divided by the average annual rate of loss of depth due to siltation, will give the optimum interval between campaigns, although this may be very variable, especially in water courses which receive substantial sediment input following heavy rainfall or where sediment deposition is relatively localised.

The optimum interval between maintenance campaigns may vary from a few months to many years. Except where a canal serves also as a drain, the rate of siltation is usually low and hence the return period for maintenance dredging to remove sediment is long. The requirement for maintenance to remove foreign debris causing obstruction may be more frequent, although generally will be localised.

3.4.4 Rivers and watercourses

Determination of the optimum interval for river dredging should be made in the same way as that described for canals. However, for rivers the constraints of maximum and minimum bed levels are usually less severe.

In most rivers the average velocity of water flow is greater than in canals, hence the rate of sediment transport may be higher. Where the velocity is highly variable across the section, or with length, both erosion and deposition may occur. Progressive sediment deposition will result in the formation of shoals in quiescent areas of low current velocity. As a result, in natural river channels, there may be a wide variation in section loss with length. In contrast, artificial channels with approximately uniform section and flow will usually exhibit a fairly uniform pattern of siltation.

Where shoaling is highly variable with length, the interval between maintenance campaigns may be dictated by those areas where the rate of section loss is greatest. In such rivers, maintenance dredging may be best tackled by a system of relatively frequent removal of isolated shoals, accompanied by less frequent dredging of the river overall. River shoals may be seasonal, with deposition occurring under particular flow conditions, followed by subsequent erosion when river flow is high. It follows that judgement concerning the necessity of dredging should not be precipitous, because a problem apparent in one season may later disappear.

3.4.5 Ponds, lakes and reservoirs

For ponds, lakes and reservoirs, siltation may be due to the import of sediments by feeder watercourses or the accumulation of organic material, or commonly due to a combination of both. Maintenance will be necessary when either capacity or water quality have fallen to an unacceptable level.

The still waters of ponds, lakes and broads, and some canals, provide a very effective trap for organic matter and agro-chemicals washed from the land. These may cause excessive enrichment of bottom sediments, resulting in anaerobic conditions and progressive deterioration in species diversity. Dredging is necessary to restore the habitat. Usually the need for maintenance of still waters will be infrequent.

3.5 DEFINITION OF THE TASK

Adequate and proper definition of the dredging task is an essential prerequisite to selection of the most appropriate method of dredging, treatment and disposal. A procedure for defining the dredging task is described in Chapter 4.

Definition of the task leads logically to the identification of the range of available solutions, or the optimum solution. Prior to definition of the task, a large number of options may exist, from *do nothing*, through a whole range of dredging, treatment and disposal options to some action other than dredging.

3.6 ENVIRONMENTAL CONSIDERATIONS

It is essential that the consequences for the environment of dredging and disposal work are fully taken into account during each stage of planning of the work. All involved have a moral duty to minimise adverse effects and maximise the benefits for the environment, so far as is reasonable. This duty is reinforced by various laws and regulations (see Section 9.2).

There is a considerable volume of published information which addresses the environmental aspects of dredging, particularly the dredging of contaminated material, but legislation and attitudes are changing continuously and so rapidly that any reference given is soon likely to be outdated. Hence the literature should be reviewed regularly. For general guidance, see CEDA/IADC (1996), Rijkswaterstaat (1996), US Army Corps of Engineers (1985-96) and WTC (1995).

Dredging is often mistakenly perceived as an operation which will inevitably have an adverse impact on the environment. This is not true. Dredging using appropriate plant and methods can maintain, or improve, environmental habitats and features of particular interest. Indeed, dredging may be essential in order to preserve particular environmental conditions.

Dredging plant and techniques specifically designed for the dredging of contaminated sediments with minimal impact offer a range of solutions for improvement of aquatic environments damaged by past activities. However, it is important that the selected method of dredging is the most appropriate and that the operation is planned and controlled such that unnecessary adverse effects are avoided and beneficial features are retained.

The selection of a dredging method and level of control required will be dependent on the particular site characteristics, as determined through the process of task definition (Chapter 4). Dredging within or adjacent to an SSSI will clearly require more thorough planning and control than dredging in less sensitive areas and will require authorisation from English Nature (see Section 9.3). However, in virtually all locations a degree of consultation will be required, and this will vary depending on the nature of the proposed work (e.g. regular maintenance dredging or capital dredging). A list of potential consultees is provided in Box 3.2.

Box 3.2 *Environmental considerations - list of potential consultees*

Council for sports and recreation
Council for the protection of rural England
Country Landowners Association
Countryside Commission
Countryside Council for Wales
County archaeologist
Environment and Heritage Service, Northern Ireland
English Heritage
English Nature
Environment Agency, England and Wales
Internal Drainage Boards
Landowners
Local authorities
National Farmers Union
The National Trust
Navigation authorities
Royal Society for the Protection of Birds
Welsh Historic Monuments (CADW)
Wildlife Trusts

It is important during planning to assess the level of sensitivity of the site environs and the likely impact of dredging operations. Sensitivity may be determined by one or more of several factors, including plant life, fisheries, invertebrates, aesthetic quality and overall environmental value. The nature of the features of interest should influence the technique, operational practice, extent and timing of dredging (Chapter 9). It may be necessary to assess the impacts of a range of dredging, treatment and disposal methods in an iterative manner in order to determine that which is most appropriate to meet the overall objectives and requirements of the site. At every stage it is important to understand the potential short- and long-term effects of dredging.

If it is necessary to remove contaminated sediments by dredging, it is important to limit contact between the dredged sediments and surrounding aquatic and land environment. It is inevitable that some sediment will be spilt, or suspended into the surrounding water, but subject to the use of an appropriate dredging method and good control, it is possible for the level of transfer of contamination to be contained within acceptable limits.

The spread of material might be restricted by the use of silt curtains (see Figure 3.3). Silt curtains create a barrier of fabric or a rising air/water mixture. Fabric curtains comprise a pervious, but fine mesh geotextile, weighted at the bottom and suspended from floating booms. Individual curtains are arranged and anchored so as to completely surround the immediate working area of the dredger. Air curtains comprise a bottom mounted pipe fed with compressed air. Air is released via small holes and air bubbles rise rapidly to the water surface, creating a vertical current of air and water through which fine sediments amy not easily pass.

The effectiveness of silt curtains in restricting the spread of suspended sediments is very dependent on the local environment and the method of dredging. Simple silt curtains are unlikely to be effective in moving water (if the current velocity is greater than 0.01 m/sec). Even when used in relatively still water, if the dredging method involves transport by hopper barge, passage of the hopper through the curtain is likely to result in the release of some suspended sediment. If possible therefore, the hopper should be positioned outside of the curtained enclosure.

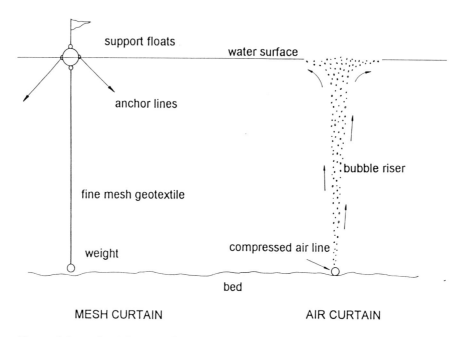

Figure 3.3 *Containment of suspended solids*

3.7 ARCHAEOLOGY

Water and sediment often obscure important relics of the past. Consent to dredge is likely to be necessary in any area of archaeological interest. Special care will be needed when dredging in such areas. A pre-dredging survey of the areas to be dredged may be required by the relevant heritage authority.

Items of archaeological interest may be found in any type of inland waterway. Obvious sites include:

- moats
- broads
- man made lakes, or natural lakes with a raised water level
- reservoirs
- rivers passing through areas of past habitation.

Items of interest may be fixed, as in the case of submerged structures and foundations, or may be loose, in the form of artefacts, weapons, coinage, jewellery, etc. Loose items may be found beneath any body of water, or marsh of significant age, but are most likely to be found in the vicinity of ancient river crossings.

Enquiries should be routinely made to the appropriate heritage authority during the planning of all dredging works. These are:

England	English Heritage
Northern Ireland	Environment and Heritage Service
Wales	Welsh Historic Monuments (CADW)

3.8 DREDGING, TREATMENT AND DISPOSAL

This report deals primarily with dredging, but the dredging process cannot be viewed in isolation. Treatment, where appropriate, and disposal must be considered simultaneously if optimum and acceptable solutions are to be found. Thus, whilst disposal is dealt with in detail elsewhere (CIRIA, 1996a), some aspects are covered in this report. For example, if dredging is by floating cutter suction dredger, which discharges via a pipeline, then the operations of dredging and disposal are a single continuous process and must be considered as such. Similarly, treatment may be an essential part of either the dredging or the disposal process.

Despite the potential for linkage, dredging and treatment are dealt with separately in Chapters 5 and 6, because although each is linked to and influenced by the other, each usually will involve separate plant and methods. Consideration of the combined processes is addressed in the description of the selection of a dredging method outlined in Section 3.11.

3.9 CONTRACT VERSUS DIRECT LABOUR

Dredging works may be carried out by Direct Labour, or by Contract. Direct Labour involves execution by the staff and employees of a public authority, usually using plant owned by the authority, however, on occasion hired plant is used.

Once the dredging task has been fully defined and an optimum dredging method or methods identified, it must be determined whether the works are to be undertaken by a Contractor, or by Direct Labour. Recently, there has been a trend away from direct labour methods in favour of contract dredging. The change has been strongly influenced by political pressure, but economic reality has also played a part. That is not to say that contract dredging is inevitably the least expensive solution. A variety of factors, including the scale and frequency of the work; the efficiency of the Direct Labour organisation; and the availability of suitable plant and skills from local contractors, will all influence the optimum method.

There is no infallible rule which determines whether execution by Contract or Direct Labour is the optimum arrangement. However, it is often the case that where the need for dredging is infrequent, Contract dredging may provide the least expensive solution, whereas when the need for dredging is continuous, Direct Labour may be more appropriate. The reasons for this include:

> - infrequent dredging by Direct Labour may fail to fully utilise plant and labour, a situation which will usually result in high unit costs
> - low plant utilisation may deprive labour of the opportunity to fully develop skills and experience
> - opportunities within the public sector to reward initiative and high productivity may be fewer than exist within the private sector.

If execution is by Contract, staff of the authority, or consultants acting on behalf of the authority, must prepare contract documents which specify the work to be done and the terms of engagement. The work should be priced by one or more specialist contractors and carried out by that Contractor which the authority deems most competitive and appropriate for the task.

Where work is carried out exclusively by Direct Labour, the selection of dredging method may be strongly influenced by the availability of particular types of plant within the organisation.

There are many arguments in the debate concerning Contract versus Direct Labour, but these are not the main subject of this report. What is important is that a realistic and unbiased assessment of the relative merits and cost of each be made in order to arrive at a valid comparison. To achieve a valid comparison it is important to take full account of financial charges (amortisation and interest on capital) and overhead costs.

3.10 SPECIALIST INPUT AND TRAINING

Dredging is a specialised activity, but the level of experience and knowledge necessary to achieve efficient operation is very variable, being strongly influenced by the nature of the site and the dredging methods employed.

For example, a hydraulic backhoe machine dredging to maintain man-made channels through farmed fen land and disposing by spreading on adjoining pasture may not require elaborate planning, nor particularly skilled operation, nor have any lasting environmental impact. In contrast, a cutter suction dredger removing contaminated sediments and discharging to a purpose-built containment site via a long pipeline may require very experienced assessment, planning, execution and supervision.

Between these two extremes are to be found projects of infinitely variable complexity. The project promoter must decide whether the task falls within his or her range of experience and expertise and whether those to be employed in carrying out the task have all of the necessary skills. If not, specialist input, or training, is advisable.

Specialist input may range from informal discussions with suitably experienced contractors to the employment of specialist consultants. If opting for the former, it should be recognised that advice given by contractors may be biased, or based on narrow experience. If employing consultants, care is necessary to determine that the firm employed has knowledge which is appropriate. Many consultants operate as general practitioners and do not have appropriate specialised experience in-house. Information concerning specialist consultants and contractors who specialise in dredging is available from the Central Dredging Association (CEDA), whose Secretariat is at the Institution of Civil Engineers (ICE), Great George Street, Westminster.

The need for training and the level of training which is most appropriate is dependent on the nature of the task and the in-house knowledge and skills of the project promoter. Clearly, plant operators should be adequately trained; such training may be available from the plant supplier. Supervisory staff, should have, as a minimum, an adequate working knowledge of relevant regulations and legislation, which may be gained by in house or external training programmes. These issues are addressed in Chapters 7, 8 and 9.

3.11 SELECTION OF AN APPROPRIATE DREDGING METHOD

There are a multitude of factors which may influence the selection of the most appropriate method of dredging. The importance of each will vary from site to site. On analysis, it may emerge that a number of different methods are suitable, or that no standard solution is practicable or acceptable. In the latter case, a special solution must be devised.

The range of options, or optimum choice of method to be employed, will emerge through the process of task definition, as described in Chapter 4. A rapid route for preliminary selection is provided by a series of flow charts, Figures 3.4 to 3.8. These identify the primary factors which influence the choice of method and guide the user simply and quickly through the available options and key point decisions, which will determine choice of method. However, many situations are too complex to be clearly represented on a flow chart. Reference should then be made to the relevant text to more thoroughly assess key decisions.

To provide assistance in preliminary selection and assessment, general guidance on the usual capabilities of common dredging plant is provided in Table 3.1. It is important to recognise that significant variations may occur between machines of different makes, or machines which are specially modified. Availability may also vary significantly (cutter-suction dredgers, for example, are scarce in some regions). The figures provided in Table 3.1 should, therefore, only be used in preliminary assessment. Manufactures, or specialist contractors, should be consulted before final decisions are made, particularly when dimensions in Table 3.1 indicate that size may be critical. For example, if height is close to the maximum which enables access beneath a particular bridge.

The flow charts list four basic matters which influence the selection of a dredging method (Figures 3.4 to 3.7). The fifth chart provides an overview and may be used first to arrive at a provisional choice (Figure 3.8).

The most appropriate order in which the four basic matters should be considered is debatable. Environmentalists have argued that environmental matters should be considered first. Others have argued that it is the dimensions of the access and work which should be considered first. However, it is the consensus view of the Steering Group, that the options for the disposal of the material usually have the greatest influence on the selection of a dredging method, and hence this is addressed by the first chart.

In fact, it is possible to start with any chart. An iterative process of chart by chart re-assessment may be necessary to reach the optimum solution.

Use of the charts requires knowledge of the conditions under which the dredging is to be carried out. These conditions, which define the dredging task, are discussed in Chapter 4.

Table 3.1 Guide to typical dredging plant capabilities

DREDGER TYPE	Size	Machine width	Height	Out-reach Standard	Out-reach Special	Down-reach Standard	Down-reach Special	Discharge distance	Discharge height	Digging capability	Production rate max.	Production rate normal
Land based	tonnes	m	m	m	m	m	m	m	m		m³/hr	m³/hr
Dragline - all rope	23	3.5	3.3	11.0	14.0	5.0	6.0	12.0	6.0	medium	103	64
	35	4.6	3.3	13.0	16.0	8.0	10.0	13.0	7.5	medium	165	104
	50	4.6	3.3	14.0	23.5	8.0	15.5	14.0	11.0	medium	225	135
Dragline - hydraulic-rope	22	3.3	3.1	15.0	15.5	6.0	9.0	13.0	6.0	soft-med	103	60
	35	3.6	3.7	15.5	20.0	6.0	8.3	15.0	9.0	soft-med	165	102
	66	4.2	4.2	20.0	25.0	9.0	11.0	17.0	12.0	soft-med	235	150
Grab - rope operated	23	3.5	3.3	9.0	12.0	10.0	15.0	10.0	5.5	medium	90	60
	35	4.6	3.3	12.0	14.0	15.0	20.0	12.0	7.0	medium	110	70
Grab - hydraulic	5	1.9	2.5	5.0	6.5	6.0	7.5	5.0	3.0	soft-med	30	15
	12	2.7	3.0	8.0	9.0	8.5	9.5	8.0	5.0	medium	50	30
	22	3.3	3.0	8.0	10.0	10.0	12.0	8.0	7.0	medium	120	80
Hydraulic backhoe	3	1.5	2.3	4.0	6.0	2.1	3.8	2.5	2.5	soft-med	20	12
	5	1.9	2.5	5.2	8.0	3.7	5.0	4.5	4.0	soft-med	30	20
	7	2.3	2.5	6.8	9.0	4.6	6.0	8.0	6.0	soft-med	40	25
	12	2.7	3.0	8.5	14.0	5.9	10.0	12.0	9.5	medium	60	40
	22	3.3	3.0	9.0	15.0	6.5	12.0	13.0	11.0	med-hard	100	60
	25	3.3	3.2	10.6	18.0	7.8	14.0	16.0	13.0	med-hard	120	70
	30	3.4	3.5	11.8	20.0	8.2	15.5	18.0	15.0	med-hard	150	90
Floating	mm	overall	air/water			Dredge depth						
Cutter-suction - auger	150	2.5	2.1/0.5	1.0	2.0	3.0	5.0	700	15.0	soft	50	30
	200	2.8	2.1/0.6	1.0	3.0	5.0	5.0	1000	20.0	soft	90	60
	250	4.0	3.3/0.8	1.0	3.0	5.0	7.0	1250	25.0	soft	190	125
	300	6.0	3.3/0.9	1.5	3.0	10.0	12.0	1500	30.0	soft	250	175
Cutter-suction - crown	200	4.0	3.2/0.9	2.0	3.0	6.0	7.0	1000	20.0	soft-med	75	50
	250	4.0	3.3/1.0	2.5	4.0	9.0	11.0	1250	25.0	soft-med	115	85
	300	6.4	3.3/1.0	4.0	5.0	10.0	12.0	1500	30.0	soft-med	210	150
	litres											
Bucket ladder	150	5.5	7.0/1.4	1.5	2.0	4.0	6.0	3.0	1.5	medium	190	120
	250	6.0	8.3/1.5	2.0	2.5	5.0	8.0	3.0	2.0	med-hard	300	200
	350	7.3	9.0/1.6	2.5	3.0	8.0	12.0	4.0	2.0	med-hard	450	290
	500	9.0	11.2/2.0	3.0	4.0	12.0	17.0	4.0	2.5	med-hard	650	400

Notes: The dimensions given here are typical but they are only a guide; plant capabilities are diverse and in all cases the actual dimensions of specific plant should be checked.
The capabilities of floating grabs and backhoes are generally similar to those for standard land based machines.
The weight of land based long-reach machines may be up to 30% greater than that listed for standard machines.
Production rates are highly variable depending on site conditions, the figures provided here should only be used as a preliminary guide to relative rates.

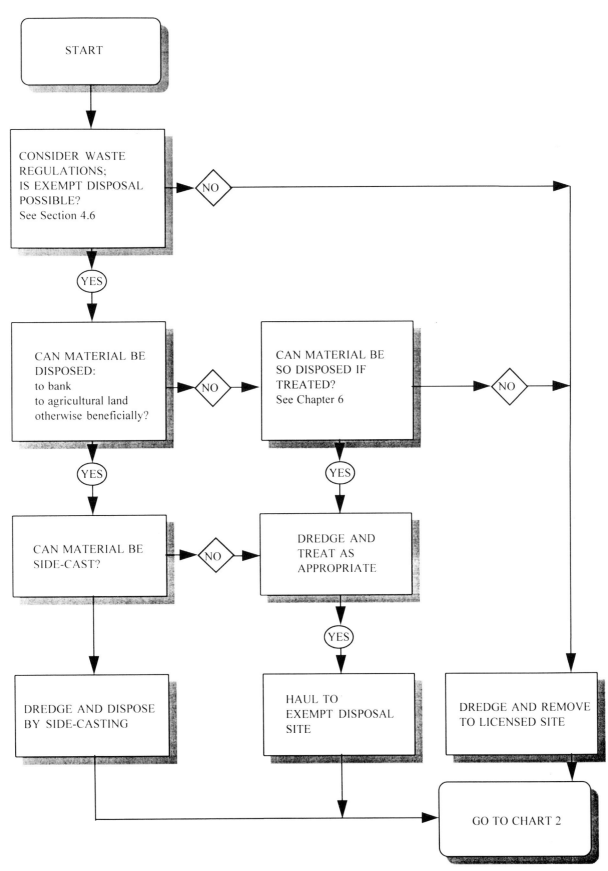

Figure 3.4 *Chart 1 - Disposal*

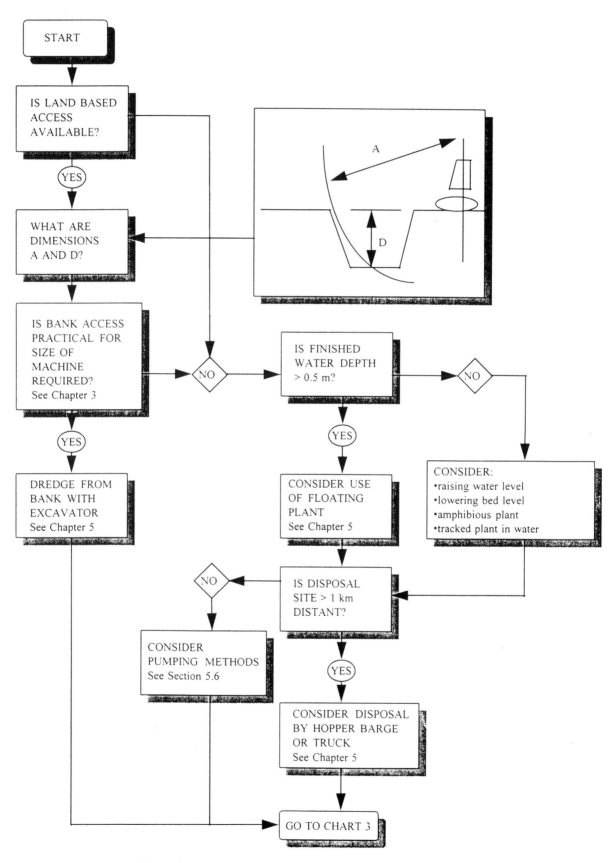

Figure 3.5 *Chart 2 - Dimensions and Access*

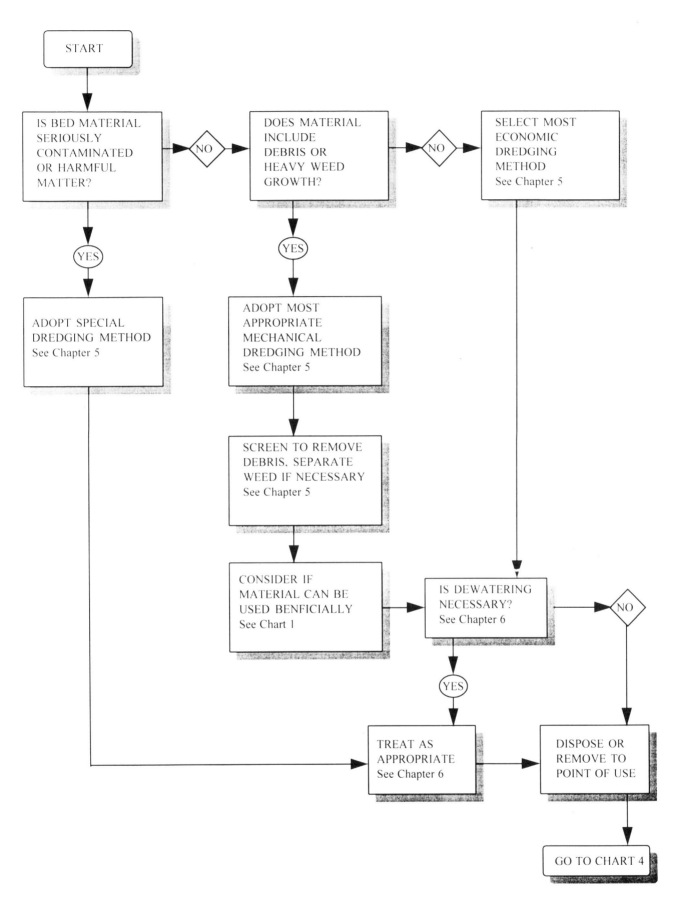

Figure 3.6   *Chart 3 - Material*

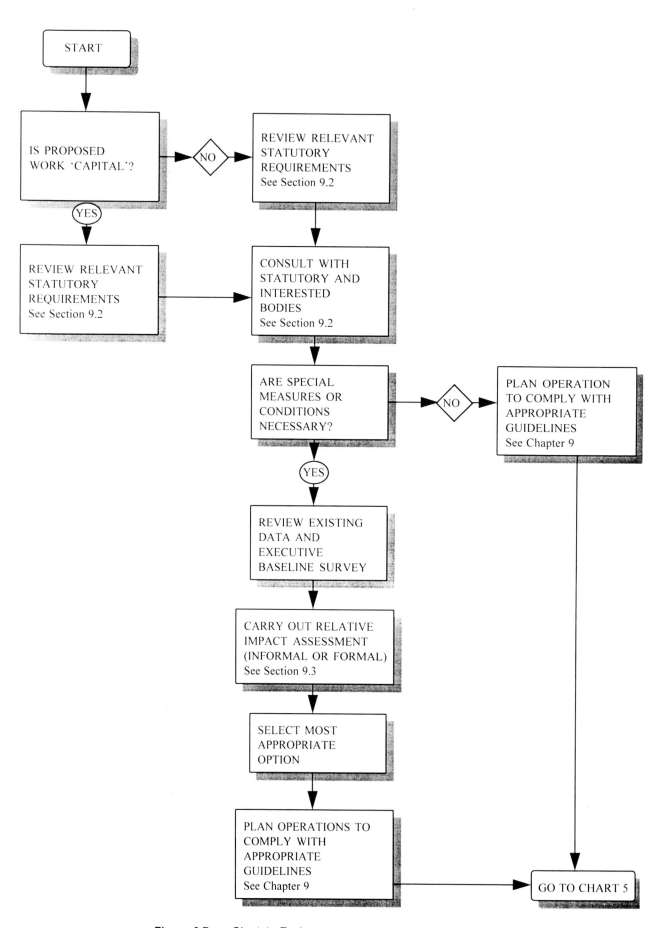

Figure 3.7 *Chart 4 - Environment*

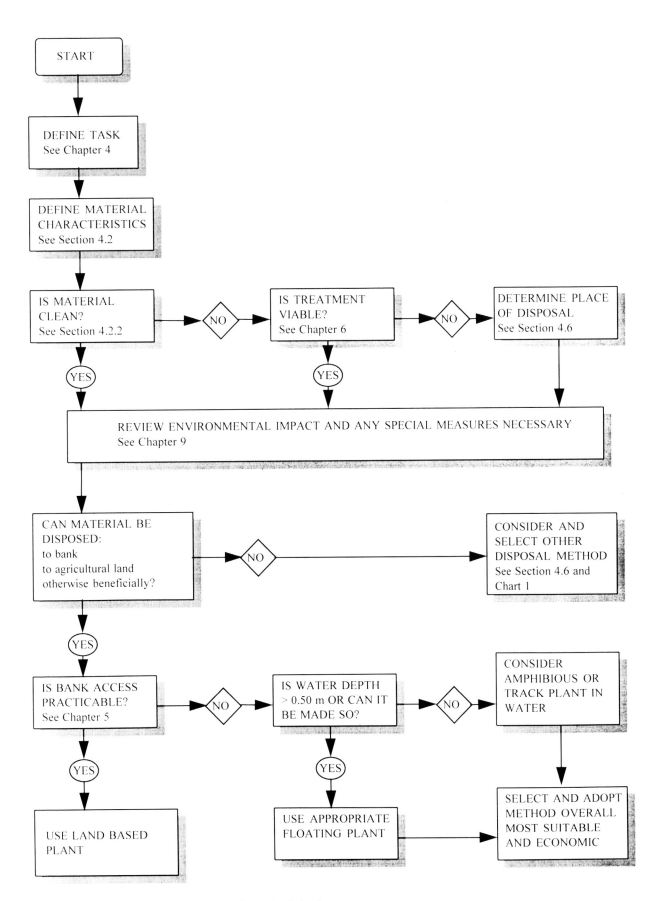

Figure 3.8 *Chart 5 - Selection summary*

4 Task definition

A proper understanding of the task to be performed and the circumstances and restraints under which it is to be performed, is an essential pre-requisite to the selection of the method which is likely to provide the optimum balance between practicality, environmental impact and economics.

The overall dredging task will usually be defined by a variety of factors. These will include some or all of the factors listed in Box 4.1 and described in the following sections.

Box 4.1 *Principal factors which define the dredging task*

• access	• sediment characteristics
• debris	• contamination
• vegetation	• environment
• season	• disposal
• treatment	• dimensions
• quantities	• security

4.1 ACCESS

4.1.1 Introduction

Access may be a critical element in selection of the optimum dredging solution. Canals in particular often occupy a narrow corridor of land with a tow path only on one bank (see Figure 4.1). Access by land-based dredging or transport plant may be restricted, perhaps severely. Bridges may be low with further restriction in tow path width, whilst tunnels may have access only for floating plant (see Figures 4.2 and 4.3). In urban areas, both canals and rivers may suffer severely restricted access. Ponds and lakes may be surrounded by delicate vegetation, or by landscaped grounds, within which disturbance by heavy plant is undesirable.

It is, therefore, essential at the planning stage of dredging work to identify the physical limitations which may restrict access. The factors which may restrict access for various types of plant are described in the following sections for both overland and floating access. In each section a check list of features to be considered is provided.

Figure 4.1   *Limited tow path access*

Figure 4.2   *Access restrictions in tunnels*

Figure 4.3 *Low bridge*

In general factors to check include:

- roads
- services
- ground conditions
- arable land
- width
- headroom
- weight
- length
- draught
- currents.

4.1.2   Overland

Metalled roads

Public metalled roads are subject to restrictions in use by earth-moving equipment, unless the equipment is suitability modified and plated (registered) for road use. The movements of track-laying plant may be prohibited and unlicensed rubber-tyred dump trucks may be limited to very short journeys. Vehicles which are not able to exceed 20 mph must display a yellow flashing light.

When the use of public roads is permitted, special care will be necessary to avoid spillage which may create hazardous surface conditions for other users. Wet material may only be transported in suitable trucks with a sealed load area. To avoid spillage it is usual to restrict loading to approximately 50% of the maximum cargo volume. As a consequence, high-sided bodies are commonly used in order to achieve an economic load.

Factors to check include:

> - traffic regulations
> - road regulations
> - consents required
> - traffic volume
> - sight lines at exit/entry points
> - surface conditions
> - suitability of proposed trucks or dumpers.

Services

Services, which include power, gas, telephone and water, whether underground, or overhead (see Section 4.1.2, **Headroom**), must be identified and considered. Underground services may need protection if they are in shallow or weak ground, in areas to be traversed by dredging, transport, or lifting plant. Special care will be necessary if working close to services which cross the waterway, especially if these are hidden below the bed. In addition to visual inspection of the site, each utility company should be requested to provide details of local services. These may include cable companies additional to conventional telecommunications companies, for example, cable TV.

Factors to check include:

> - electricity
> - gas and oil
> - water
> - telephone (all providers)
> - television (all providers)
> - other communications.

Ground conditions

The strength of soft or water-logged ground is unlikely to be sufficient to support the weight and frequent passage of trucks or dumpers. If very weak, the ground may not support track-laying plant (see Figure 4.4). If no alternative access is feasible, it may be necessary to construct a temporary access using geotextiles with granular topping. Re-usable and portable flexible metal surfacing may provide solutions over short distances, but may be expensive.

Weak or sensitive ground, such as lawn, may also render access difficult for floating plant which is delivered to the work site by road. For example, pond or lake restoration often will employ floating cutter suction dredging plant which is transported by road vehicle and off-loaded and launched by mobile cranes. Axle loads of transporter and cranes may be very high, requiring substantial advance work to protect sensitive paths and grounds which typically surround ornamental waters.

Factors to check include:

- general strength of ground - seasonal variation
- bank edge stability
- sensitivity (lawns, etc.)
- land owner consent.

Figure 4.4   *Weak ground conditions*

Arable land

Arable crops are commonly grown within one or two metres of river and lake banks. Usually impact can be minimised by programming work such that access during the crop growing season is avoided. Nevertheless, the land owner will usually wish to follow cropping quite quickly with re-cultivation and hence the window of opportunity for access may be short. If crop damage or disruption of the cultivation cycle cannot be avoided, payment of compensation may be necessary. Running over arable ground may cause damage to the soil structure through excessive compaction, or result in the ground turning to slurry in wet conditions.

Factors to check include:

- crop cycle
- land owner consent.

Width

It may be impossible to overcome restrictions to the width of access in a cost effective manner, particularly when the restriction is caused by structures (see Figure 4.5). Plant must then be chosen which is able to operate within the available space. Accurate determination of the width may be critical in both the selection of plant and overall planning of the work. It is important that the entire access route be inspected and any width restrictions be carefully measured. In addition, it should be borne in mind that the full width of ground may not be available, for example, where the bank edge is too weak to support a substantial load. Where restrictions are severe, land access may not be feasible.

Factors to check include:

- minimum width between fixed features (buildings, structures or trees)
- gateways
- bank edge stability.

Headroom

Similar comments apply to headroom as for width (see Figure 4.1 and 4.2). In addition to structures, trees and other vegetation, headroom may be restricted by utilities, such as power and telephone cables. Clearly, special care and precautions are essential when operations are planned in the vicinity of live overhead power cables. Height restrictors with audible or visual alarms should be erected on the approaches to power cable crossings. The power company will give guidance concerning safe operating distances and stipulate appropriate precautionary measures.

Figure 4.5   *Restricted width*

Factors to check include:

- clearance beneath structures
- safety regulations
- safe working height beneath power lines
- clearance beneath telephone lines.

Weight

Consideration must be given to possible adverse loading due to the weight of dredging, transport, delivery and lifting plant. Damage may otherwise be caused to bridges (see Figure 4.6), culverts, pipes, cables, surfaces, ground, etc.

Factors to check include:

- bridges
- culverts
- underground services
- underground structures (cellars, reservoirs, etc.)
- road or path surfacing
- ground conditions.

Figure 4.6 *Weak access bridge*

4.1.3 Floating

Width

The submerged cross-section of floating plant may be rectangular, as in the case of pontoons, whereas dredged canals and rivers will usually have a cross section which is approximately trapezoidal. Hence the critical width will usually be that measured at the depth of maximum draught of the proposed floating plant relative to the minimum anticipated water level.

When loading dredged material into independent hopper barges, with most mechanical dredging plant it is necessary for the barge to be alongside the dredger during loading, in which case the combined width of dredger and barge must be accommodated (see Figure 4.7).

If it is necessary to preserve access for other traffic during the dredging operation, then additional width for the beam and safe clearance of passing vessels is required.

The overall operational width of floating plant may be greater than the width of the vessel structure. Width may be increased by fendering, fixed fairleads, spud carriers, etc.

Factors to check include:

- channel water width at maximum draught of anticipated plant
- operational width of proposed plant, or combined plant
- width at structures
- width to accommodate passing traffic
- approvals from navigation authorities.

Length

Length must be considered in several respects. Travelling distance must be determined for floating plant, both in relation to the initial mobilisation and also in relation to the proposed method and place of disposal. It may also be necessary to determine the distance to a suitable turning area if the plant to be employed must be turned, as will be the case for self-propelled hoppers. The length of the hoppers will determine the space needed for turning.

Factors to check include:

- travel distance from launch point (mob. and demob.)
- travel distance to disposal point relative to work site
- travel distance to turning area, if required
- length of hopper relative to turning area
- length of channel to be dredged.

Figure 4.7 *Obstruction to navigation caused by dredger and hopper barge*

Headroom

Floating dredging plant may require greater headroom than is required by normal traffic. The overall air section (cross section above water) of floating dredging plant most probably will also be greater. It is, therefore, important to consider both the height and width above water (see also Section 4.1.3, **Width**). The general considerations are similar to those for land based plant, but where bridge or tunnel air draught, or section, is too small to allow passage for floating plant, the cost to by-pass the obstruction may be much greater than for land-based plant. In most cases it will be necessary to remove floating plant from the waterway and re-launch, perhaps involving dismantling and re-assembly, beyond the obstruction. A few types of plant are able to manoeuvre around structures (see Figure 4.5).

Factors to check include:

> - air draught and section
> - safe working height beneath power lines.

Draught

It will usually be the case that the dredging plant to be employed will be designed to operate with a similar draught to that required by normal waterway traffic. However, where the channel section has suffered serious deterioration, normal water depth, or section, may not be available. Greater depth will be created as dredging progresses, but access may be limited to the newly dredged length. When water level is subject to seasonal variation, it may be necessary to programme work to coincide with the most advantageous water levels.

The draught requirements of floating plant cannot be determined until the plant to be used is decided upon, which may not be known at the time of task definition. However, the existing available water depth and that available on completion of dredging can be determined given knowledge of existing and proposed bed levels and water levels. <u>Where water level is variable, it is the minimum level which should be considered.</u>

Clearly the water draught of all floating plant will increase when loaded. However, in the case of hopper barges, the weight of a full load may vary. A hopper fully loaded with sand will have a deeper draught than if filled with mud or vegetation. It may not be possible to carry a full load of sand. Maximum draught will also alter with trim; if a hopper is loaded unevenly its maximum draught may be greater. For self-propelled hoppers, draught at the stern will usually increase when under way and if the hopper section is large relative to the water section (high blockage factor), the draught when under way may be greater than when stationary or when in open water.

Factors to check include:

- minimum and maximum water levels
- minimum and maximum bed levels
- minimum and maximum draught of plant to be employed.

Current

Strong currents are not common in UK inland waterways which are accessible to floating plant, but exceptions do exist. Strong currents may be found in large irrigation canals overseas and of course in some rivers. Current strengths may be affected in future years if potential water transfer schemes are implemented. For operational purposes, 'strong' currents may be described as in excess of 1 knot (0.50 m/sec).

Current will affect the operation of floating plant in several ways. Hopper transit times and handling may be adversely affected by strong currents. Similarly, the operation of floating dredgers may be impaired. Currents will also influence the dispersion and distribution of suspended sediments and hence may be important if potentially adverse effects on water quality, or nearby sensitive aquatic areas, are to be avoided.

Factors to check include:

- maximum current velocity and seasonal variation
- if applicable, direction of current.

Weight

Two important aspects of weight must be considered. One is the effect on draught, as previously described, the other is its relation to access, where the dredging plant must be delivered overland (see **Weight** considerations for land based plant in Section 4.1.2).

Factors to check include:

> - maximum axle loading of transporter
> - maximum axle loading of cranes
> - maximum load imposed by cranes during lifting
> - effect on draught.

Towing

Where the dredging plant is to be towed to the site via the waterway network, it is necessary to consider the requirements of the towing vessel, which may require greater water, or air draught, than does the dredging plant. If strong currents are expected, the necessary bollard pull of the towing vessel will be greater. If towing through exposed areas, such as fen land, or water meadows, wind strength may also influence towing loads. Cross-winds may create difficult towing conditions.

Factors to check include:

> - water and air draught of proposed tow vessel
> - bollard pull in relation to expected currents and wind strength
> - closures (canals closed to navigation).

Lifting

Specialised dredging plant may be delivered overland and assembled and/or off loaded by heavy cranes. For lakes and ponds, overland access is usually the only option.

The axle loading of large cranes may be substantial, as may be the loads imposed during lifting. The required size of crane, or cranes, is determined by the maximum load to be lifted and the maximum radius of lift. The radius may also be substantial, particularly when lifting is to take place on the bank margin of a pond or lake. It may be unsafe to place the cranes close to the bank edge and the distance from the bank to water of sufficient depth for the plant to float will most probably be significant. When launching a dredger using cranes standing on an unsupported bank, the necessary operating radius is likely to be in the range 7 m to 10 m, but may be greater.

The size of crane required to safely handle the load at extended radius may increase greatly relative to that required to lift at minimum radius. This is particularly true of cranes with telescopic jibs. Radius is measured from the centre of rotation of the crane to the centre of the load to be lifted, NOT from the extended outriggers, or edge of tracks, or load. For large loads at extended radius it may be necessary to use two cranes lifting in tandem (See Figure 4.8). Such operations require considerable skill on the part of the crane operators and should not be entrusted to inexperienced operators (see BS 7121). Table 4.1 provides a guide to the relationship between safe lifting capacity and radius for typical modern mobile cranes with telescopic jibs.

WARNING: Heavy lifting is a specialised activity, especially so when cranes are to lift in tandem. <u>Specialist advice should always be obtained</u>.

Table 4.1 *Typical lifting capacities for modern mobile telescopic cranes*

Rated capacity in tonnes	Radius of lift measures	Safe load in tonnes
30	3.0	30
30	5.5	15
30	11.0	5
60	2.5	60
60	3.5	50
60	6.0	30
60	10.0	15
60	15.0	7
80	2.5	80
80	5.0	40
80	10.0	20
80	15.0	10

Note: Figures are for 75% tipping load, but are approximate and may vary between cranes. All capacities assume outriggers fully extended and 360° operation. Lifting capacities will be reduced if jib length is greater than minimum necessary for radius.

Factors to check include:

- maximum load to be lifted
- maximum radius of lift
- crane size required
- maximum axle loading
- surcharge imposed on bank or quay during lifting.

Figure 4.8 *Tandem crane lift*

4.2 SEDIMENT CHARACTERISTICS

It is important to sample and analyse the bed material to at least the full depth which is to be dredged, particularly when capital dredging is being planned. The range of strengths of sediment to be dredged in maintenance dredging, being of recent fluvial origin, will usually be far less than in capital dredging, where the material to be dredged may range from soft mud to hard rock.

The physical characteristics of ground or sediment will have an important effect on the performance of dredging plant and on the aquatic environment in the area of dredging operations. The chemical and biological properties of the sediment may influence the method of dredging and will influence the method of disposal and the potential for beneficial use.

When it is anticipated that treatment of the dredged material may be necessary, it is important that the programme of sampling and analysis provides the information necessary for the assessment of the optimum treatment process and an estimation of the costs.

4.2.1 Physical characteristics

Capital dredging

Capital dredging may require the removal of a wide variety of material, possibly ranging from soft mud to rock. Between these extremes there are many different types of material, including clays, sands, gravels, peat, rock of varying strength, etc. Each may require different methods and some will only be dredged at much reduced rates of production and hence greatly increased cost.

Recommended methods of investigation, sampling and testing in relation to dredging work are described in BS 6349, Part 5, 1991.

The procedures for investigation and testing are described in BS 5930, 1981 and BS 1377, 1990 respectively.

Generally it is the variable strength, particle size and moisture content of soils which will have the most important effect on dredging and disposal. Contaminant level, if any, will have an influence on the dredging method and be critical for treatment and disposal. If material is believed to be contaminated the particle size distribution should be determined down to a size of one micron (see **Maintenance dredging** and Section 4.2.2).

Factors to check include:

> - *in situ* strength
> - *in situ* density
> - *in situ* moisture content
> - particle size distribution.

Maintenance dredging

The potential variation in the physical properties of sediment to be dredged for maintenance purposes is less than for capital dredging because only recent fluvial deposits must be removed, but significant variations in particle size may occur. These may range from clay to gravel sized particles (0.0001 mm to 0.063 mm). Particle size will influence the performance of dredging plant, particularly those types which raise and transport soil hydraulically. Particle size may also influence the potential for beneficial use. A wider range of useful applications exist for sands and gravels compared with mud (see PIANC, 1990). High moisture content, which is a common characteristic of sediments dredged for maintenance purposes, may render transport and disposal more difficult and costly.

Factors to check include:

- in-situ density
- in-situ moisture content
- particle size distribution
- organic content.

4.2.2 Chemical and biological characteristics

Bottom sediment in waterways which pass through areas of past or current industrial activity may be polluted by chemical, organic or heavy metal wastes which have been discharged or spilled to the waterway. Even in non-industrial areas, contamination may be caused by the natural leaching of dissolved elements from rocks and ground and subsequent adsorption or absorption onto sediment particles. In agricultural areas sediments may become contaminated by fertilisers and pesticides. Discharges from sewage works, sewers, or from rural properties without the benefit of mains drainage, may also cause enrichment of the bed sediment.

The chemical and biological properties of the material to be dredged may influence the method of dredging and disposal and may dictate whether treatment is required. If the bed material is contaminated by substances which are potentially harmful, whether due to toxicity, or increased oxygen demand, special care and dredging methods may be necessary to avoid harm to the aquatic or surrounding environment. Usually it will be necessary to adopt dredging methods which minimise the release and spread of potentially harmful matter.

The character of the bed material is determined by the recovery and testing of samples. In the case of maintenance dredging, unless the depth to be removed exceeds 500 mm, it will usually be sufficient to recover a near surface sample by means of a hand operated sampling grab, or if convenient, by using a grab or backhoe dredger. When capital dredging is planned, or where the depth to be removed exceeds 500 mm, an intrusive method of investigation, such as boreholes, or vibrocores, may be necessary. Routine sampling and testing will not usually be necessary for inland waters with no history of pollution. Material arising from the maintenance of drainage channels in agricultural areas may be enriched by fertiliser laden run-off, but such material is unlikely to be harmful and indeed may be beneficial if returned to the land.

Contamination may bar dredged material from beneficial use and complicate other disposal options. It is, therefore, essential to sample and test materials to be dredged in a proper and representative manner. The range of potential pollutants is extensive and if doubt exists concerning which tests should be undertaken, advice should be obtained from a specialist, or from the Environment Agency. Pollutants to test for are described in the check list which follows. Disposal is discussed in Section 4.6. For detailed information concerning testing and sediment classification see CIRIA (1996a).

At the time of writing (June 1996) there is no national system for the classification of dredged material. A draft consultation document, *Development of a National Waste Classification Scheme Stage 2: a system for classifying wastes*, was published by the DOE in December 1995.

Material samples recovered should be tested in an accredited soils testing laboratory. Separate leachate tests may also be necessary. Where appropriate, laboratory analysis should check for:

- heavy metals
- arsenic
- pH
- sulphide
- phenols
- phosphorus
- oil
- PCBs (polychlorinated biophenols)
- TCBs (trichlorobenzene)
- PAHs (polycyclic aromatic hydrocarbons)
- bacteria.

The selection of tests from or in addition to the above list should be based on the characteristics of the dredging site and the intended method of disposal. The cost of laboratory analysis for a common suite of tests is approximately £200. If dredging is to remove only granular deposits from an upland river, it may be that none of the listed chemical or biological tests are necessary. In contrast, for material arising from a river or canal which drains areas of past or present heavy industry, all of the tests may be necessary.

A range of factors will influence whether or not testing is necessary and, when it is necessary, those tests which should be made. Testing will be influenced by the history and character of the waterway, the proposed method of dredging, the intended end use for dredged material, if any, and the proposed methods of treatment and disposal.

4.2.3 Pollution by debris

The incidence of debris varies greatly according to location, but is most common when dredging waterways and ponds in urban areas. If found in significant concentration, debris may cause considerable difficulty, especially if suction dredging methods are employed.

There is no reliable established scientific method, or procedure, by which the degree of pollution by debris may be characterised. It may be useful in navigable waters to conduct a magnetometer survey to locate substantial submerged or buried metal objects, or concentrations of small metal objects. However, the method is not precise and will only detect ferrous metals. The equipment and survey results also require skilled use and interpretation.

For some rivers, canals, ponds and shallow lakes, it may be possible to observe the bed during clear water and bright sunlight conditions, particularly if water levels are low. It may also be possible to lower water levels over a length of canal for a short period to facilitate inspection (see Figure 4.9). If such methods are not practical it may be necessary to conduct a survey by diver, trial grabbing, or raking.

It is the experience of contractors who are regularly engaged in the dredging of inland waterways, that where inspection reveals little or no debris, some is to be expected, and where some debris is visible, much more is to be expected (see Figure 4.10).

Whichever method of assessment is employed, the objective should be to determine approximately the density and size range of debris. It may not be economic to survey long lengths of waterway, but it should be possible to select an appropriate number of representative, but short lengths, perhaps 25 m each, and to conduct a detailed count and analysis within each representative length. It is sensible to concentrate on those places with easy public access such as bridges and in urban areas. The results should illustrate the approximate density and size distribution of foreign matter. Size distribution might range from house bricks to abandoned vehicles and this will determine whether special and expensive methods of removal will be necessary.

If it is impracticable to measure and describe the concentration of debris and if contract dredging is planned, the minimum which should be done is to formally draw tenderers attention to the possibility of debris. If the local knowledge of experienced staff is available, it will be helpful to provide an experienced assessment of the level of risk.

Figure 4.9 *Assessment of debris in drained canal*

Figure 4.10 *Assorted debris recovered from inland waterway*

Approaches to consider include:

- undertaking a visual check for debris (almost certainly present in urban areas)
- describing the distribution for typical lengths
- conducting a diver, raking, or grab survey where debris is suspected, but not visible.

4.3 VEGETATION

Some types of vegetation may seriously impede dredging work, particularly when suction dredging methods are employed. Any dense concentration of water weed is likely to cause significant problems for suction dredging. When assessing the work to be done, weed growth, or other growth, such as alder carr, should be described in terms of the type of vegetation, percentage coverage and whether it is to be removed or preserved.

WARNING: Aquatic vegetation may be important for various reasons and may need to be preserved or protected. Specialist advice should be sought prior to its removal.

Approaches to consider include:

- identifying types of vegetation, if significant
- estimating percentage coverage
- differentiating between areas for preservation and removal.

4.4 ENVIRONMENTAL CONSIDERATIONS

In this report, the environment is considered only in relation to the impact which dredging or disposal might have upon it. Broader issues concerning the overall environment and ecology of inland waterways are not addressed. Readers are referred to *The New Rivers and Wildlife Handbook* (RSPB, 1994).

Most waterways will support a variety of flora and fauna. Usually it will be necessary to minimise the effect of dredging work and in particular, to avoid, or minimise adverse effects. For example, in waters which support a significant fish population it may be necessary to avoid an excessive increase in suspended solids. Aquatic and bank-side vegetation may be important for different reasons. Rare vegetation should be preserved or protected, while common and prolific vegetation may simply impede the dredging operation. It will usually be desirable to retain at least a narrow margin of vegetation to protect against bank erosion and provide cover for wildlife.

Thus the environment in which dredging is to be carried out may directly influence the method of dredging which should be selected. In order that dredging work may be properly planned and achieved with minimum harm, it is necessary to have an understanding of the local aquatic ecology. If bank-side access or disposal is planned, it may also be necessary to consider the ecology of the waterway margin by means of survey, such as a river corridor survey.

Dredging will inevitably influence the environment of a waterway and cause some degree of change. The environmental attributes of canals are by their very nature artificial, and ultimately dependent on the maintenance of the waterway. Dredging in canals should, therefore, be viewed as environmentally acceptable, provided that it is appropriately designed and implemented, taking full account of the environmental features requiring conservation or enhancement.

Improvement of the aquatic environment may be the prime reason for dredging. For example, The Broads Authority have engaged in extensive dredging works in the Norfolk Broads specifically to remove harmful bed material and improve the aquatic environment. The works have resulted in the recovery of the system and a far greater species diversity. Similar beneficial work has been undertaken throughout Europe in canals, ponds, lakes and rivers. Dredging is increasingly seen as a tool for positive use in achieving environmental improvement.

When planning a dredging operation the nature of the site should be considered and the environmental features of interest identified. These will determine the nature and extent of the dredging which would be appropriate, as well as identify any relevant seasonal factors such as fish migration, breeding or plant seeding or growth. The views of Environment Agency fisheries officers should be obtained.

If it is planned that the proposed work be facilitated by a significant reduction in normal water level, or complete de-watering, the work should be planned to minimise the adverse impact. This may involve planning the work to take place in the least sensitive season, or the temporary removal of fish stock and selected vegetation. De-watering may also adversely affect the stability of banks or structures.

The ecology of both the watercourse and banks should be evaluated in terms of the type, numbers and importance of flora and fauna. Seasonal variations should also be assessed. Numbers may be difficult to determine, especially for mobile species, but the project promoter will usually have sufficient knowledge of the local ecology to describe flora and fauna in general terms. If such information is unavailable, a survey of the site of proposed work should be made by a specialist. This may need to extend beyond the limits of proposed work; for example, when a sensitive natural or commercial site exists downstream, or within the sphere of potential influence, e.g. a fish farm. Further information is provided in Chapter 9.

Dredging carried out for or by authorities which have a duty to further conservation (e.g. Environment Agency, British Waterways) must consider environmental issues in line with that authorities Environmental Code of Practice.

A further consideration is the mandatory and non-mandatory organisations which should be consulted. In England, for works in SSSIs, it is a statutory requirement to consult English Nature; for designated historic monuments English Heritage is the statutory authority. The Environment Agency has responsibility for the protection of water courses in England and Wales and are an obvious consultee for environmental issues. A range of local organisations with environmental interests should also be consulted (see Box 3.2).

Approaches to consider include:

- describing all important flora, its distribution and abundance
- describing all important fauna, its distribution abundance and the breeding requirements relevant to the site
- describing sensitive areas within sphere of influence
- the presence of geomorphological features, such as riffles.

4.5 SEASONAL FACTORS

There are many seasonal factors which may restrict working time:

- pleasure craft traffic on narrow canals is usually greatest in summer
- width may be inadequate for dredging equipment and the passage of other vessels
- it may be necessary to avoid work during peak periods
- in rivers which are host to migratory fish, it may be necessary to avoid work during the main migratory period
- work in rivers may be difficult during periods of high flow
- disposal to, or access over, farm land may only be acceptable between cropping, or in the case of pasture, when free of stock
- open and slow moving water, such as lakes, broads and canals, may freeze in winter, rendering some types of dredging work difficult or impossible
- canals are often closed in winter for repair work
- many species are more vulnerable in the summer.

The range of seasonal factors which may exert an important influence on dredging and disposal operations is very diverse. Unless the project promoter has a thorough understanding of the range of working methods most likely to be employed it is advisable to seek specialist advice, otherwise an important factor may be over-looked, perhaps giving rise to operational problems later.

Factors to check include:

- water use
- water flows (maximum and minimum)
- water levels (maximum and minimum)
- fish spawning, migration, sport
- water fowl and other birds breeding, nesting, migrating or over-wintering
- land use - access and disposal
- growth of valuable plant life
- weather
- daylight hours
- public access.

4.6 DISPOSAL OPTIONS

The place and method of disposal of material arising from dredging will have a major influence on the overall cost and environmental impact of the works. For the dredging of inland waterways a range of disposal options may be identified. The main options are listed in Box 4.1. The list is arranged approximately in ascending order of cost and environmental impact.

Some of the methods listed in Box 4.2 are exempt disposal options under the Waste Management Licensing Regulations 1994:

- Methods 1 to 3 are exempt from licensing under the Regulations (with certain restrictions; see CIRIA (1996a)) and will require registration
- For methods 4 to 11, it is possible that an exemption may be registered, or that a Waste Management License may be required
- For methods 12 and 17 a License will be required.

If the only available method of disposal is by transport to a licensed commercial waste disposal site it is most likely that it will be necessary to reduce the sediment water content to render the material into a condition more easily handled and spread (see Chapter 6). Reducing the water content will also reduce the weight and hence the total charges imposed by the disposal site operator. However, the de-watering of contaminated sediments may produce contaminated water which must be treated or disposed of separately, or may increase the percentage level of contamination such that a higher level of treatment or more expensive disposal option must be adopted.

Commercial disposal site operators may be unwilling to guarantee acceptance of the material. This may cause major difficulty in estimating the cost of disposal. If the operator of the site closest to the works refuses to accept the material arising from dredging, or demands punitive charges, the haul distance to the nearest alternative site may be very much greater. At present, in some regions, there are few if any sites willing to accept substantial quantities of wet material.

When reviewing the options for the disposal of dredged material, it is important to recognise that there may be substantial differences in density of the material to be disposed of relative to that *in situ*. Unless de-watering treatment is carried out, the material density will usually be reduced due to loosening and increased water content. This is particularly true when pumping methods are used to dredge and transport fine grained materials which are not free draining. If material is to be pumped to a disposal lagoon, the lagoon capacity may need to be substantially greater than the *in situ* volume of dredged material which is to be accommodated. An estimation of the increase in bulk may be made by means of settling and consolidation tests in a soils laboratory.

Box 4.2 *Disposal methods listed approximately in ascending order of cost and environmental impact*

---

1. Discharge and spread directly on waterway banks (see Figure 4.11).
2. Discharge and spread thinly over adjacent agricultural land (see Figure 4.12).
3. Discharge adjacent to waterway for later use in flood bank improvement.
4. Pump discharge to nearby stockpile for subsequent beneficial use.
5. Transport by hopper barge to nearby stockpile for subsequent beneficial use (e.g. on agricultural land).
6. Transport by hopper barge to remote bank stockpile for subsequent beneficial use.
7. Pump discharge to remote stockpile for subsequent beneficial use.
8. Transport overland by dump truck to stockpile for subsequent beneficial use.
9. Transport over public roads by closed truck to stockpile for subsequent beneficial use.
10. Improve by treatment and spread on adjacent agricultural land.
11. Improve by treatment and spread on remote agricultural land.
12. Pump discharge to nearby licensed disposal site.
13. Transport by hopper barge to nearby licensed disposal site.
14. Transport by hopper barge to remote bank-side disposal site.
15. Pump discharge to remote licensed disposal site.
16. Transport overland by dump truck to licensed disposal site.
17. Transport over public roads by closed truck to licensed disposal site.

A range of parameters require definition specifically in relation to disposal. Factors to check include:

- quantity to be disposed
- dimensions of bank-side land available for disposal
- area of agricultural land available for disposal
- rate of disposal possible in cubic metres per linear metre of bank, or hectare of land, and limits on rates of disposal under the WMLRs
- proximity and dimensions of embankments to be improved
- pumping distance and relative elevation to stockpile/disposal area
- sailing distance to stockpile/disposal site
- overland haul distance to stockpile/disposal area for dump trucks
- road haul distance to stockpile/disposal area for road vehicles
- maximum stockpile capacity required
- maximum disposal site capacity required
- can the disposal method be registered for exemption?
- is treatment to allow exempt disposal viable?

With the possible exception of heavily contaminated materials, it is likely to be the case that disposal off site at a licensed disposal facility will be the most expensive. It follows that all alternatives should be thoroughly investigated. Clearly, bank-side disposal or disposal by spreading on agricultural land are likely to be the least expensive. In the latter case, it is important that no harm is caused to the agricultural properties of the land; indeed, demonstration that improvement will result is a requirement.

Figure 4.11 *Bank-side disposal*

Figure 4.12 *Disposal on agricultural land*

Other disposal methods close to the waterway should also be investigated, some of which may not be obvious. For example, existing flood embankments may be in good repair, or of adequate dimensions, but an increase in dimensions may provide the means of disposal of dredged material and simultaneously, by widening, improve the embankment as a means of access, or by slope reduction, improve the grazing for stock.

4.7 TREATMENT NEEDS

The need for treatment of dredged material may arise for a variety of reasons. Treatment is usually to render material harmless, or less harmful, or to enable simpler or more economic disposal. The simple measure of reducing water content may permit more economic disposal by rendering the material acceptable to a more convenient disposal site. Alternatively treatment may render material suitable for beneficial use, such as topsoil.

Even if dredged material is to be disposed of in the simplest way, on farm land, or land adjoining the waterway, it will usually be necessary to remove foreign matter, such as debris, or over-sized matter such as rocks and tree branches. It will be necessary to dispose of foreign matter separately.

If the material is contaminated more sophisticated treatment will be necessary, in which case, bank-side disposal is unlikely to be practicable.

Treatment techniques are discussed in Chapter 6. At the stage of task definition, it is necessary to determine only what is likely to be required.

Factors to consider include:

- is some form of treatment necessary?
- is treatment and exempt disposal cheaper than disposal to a licensed site?
- which treatment methods are likely to be required and are acceptable?
- must treatment be carried out remote from the site of dredging?

4.8 DIMENSIONS OF THE WORK

The dimensions of the proposed work will define the task in the broadest sense. Ultimately the dimensions will strongly influence the dredging method, the rate of production and the cost. When the object of dredging is to completely remove material from the waterway and relocate it elsewhere, rather than to simply relocate it within the waterway, as in shoal dispersal, the rate of production which can be achieved will depend on the volume to be removed per unit length of waterway. Usually it will be more expensive to remove one cubic metre of sediment from a bed area of 10 square metres than from an area of only one square metre. The controlling factor is how much material can be reached by the dredger from a single position. For most dredging methods, time occupied in movement of the dredger will be unproductive. Hence the thickness of material to be removed is important (see also Section 7.3.1 **Tolerances**).

The width of the work area is also important, particularly where land based plant is to be employed. If work is to be carried out with a machine standing on one bank, the distance to the far side of the area to be dredged will determine the size of machine necessary (see Chapters 5 and 7). The width may be less critical if the work is to be done with floating plant, or by use of a tracked machine standing in the waterway (see Figure 4.13). Nevertheless, in these latter cases, width will still influence production in the same way as thickness.

Bank height will also influence the performance of the dredging plant. For machines working from the top of banks the machine outreach which is possible will reduce with increasing bank height relative to bed level. For suction dredgers with pump discharge to land via pipelines, bank height will influence the hydraulic head to be overcome by the pump. Beyond a certain point, pump performance will decline with increasing head.

The product of depth, width and length will give the volume to be dredged. Some dredging works may be evaluated on the basis of volume, others on the basis of area, the method of evaluation being influenced by the type of plant to be employed for dredging, but also by the average thickness of material to be removed.

As a general rule, volumetric evaluation is likely to be preferable when the average thickness to be removed is in excess of 300 mm.

Volume often will become a crucial factor when considering disposal, particularly if the material is to be treated.

Figure 4.13 *Tracked machine working from bed in shallow water*

Factors to check include:

> - length of work area
> - width of work area (minimum, maximum and average)
> - thickness to be removed (minimum, maximum and average)
> - bank height relative to bed level and water level.

4.9 QUANTITIES

The most suitable unit of measurement of quantities will depend on the dimensions of the work (see Section 4.8). If the area to be dredged is large relative to the volume to be removed, it is likely that measurement of the work by superficial area or length of waterway, will be most appropriate. Both are relatively simple to measure. Volumetric measurement is more difficult (see Section 7.3).

Regardless of the method of measurement adopted, volume may be important when considering the options for disposal of the dredged material. In most cases it will be helpful to estimate the total volume which is likely to arise. When estimating quantities, it is important to understand that if dredging is to achieve a particular specified level, or water depth, some over-dredging will be necessary. The vertical measure of necessary over-dredging is variable, being influenced by the dredging method, the skill of the dredging plant operator and objectives of the work. In routine dredging work, over-dredging may average 150 mm, and may be more. <u>Where the average thickness above the specified level is small, over-dredging will have a very important influence on the total quantity removed.</u>

When dredging is by Contractor, whether or not over-dredging should be included in the measurement for payment is a contractual matter (see Chapter 8). Regardless of whether or not over-dredging is measured, it is work which must be done and paid for directly or indirectly.

The quantity to be transported and disposed may also be increased by bulking. This is because the volume of material after dredging may be greater than the pre-dredge volume *in situ* as a result of movement of the soil particles and filling of the increased voids by air and water (see also Section 4.6). Added water may also increase the total weight to be disposed.

Large quantities arising from dredging are often a surprise to lay people. In particular, the owners of ornamental lakes which have silted up are often surprised when advised that disposal of sediments dredged during restoration may require up to one acre of land for every acre of water cleared.

Factors to check include:

> - length, area, or volume to be dredged.

4.10 SECURITY

Two aspects of security should be considered:

- public safety
- protection of property.

The public may be at risk from the operation of dredging plant, or from the disposal of the dredged material. The actions of children, in particular, should be considered. People should be excluded from the working range of operating plant and from the area of disposal.

Property, in the form of dredging plant and equipment, may be at risk from vandalism and theft. These problems are usually most acute in urban areas, or locations in fairly close proximity to urban areas. The perceived level of risk will determine the level of security which is necessary. Plant which is to be left unattended on site may require special protection in the form of covered glass, improved locks, engine immobilisers, etc.

Factors to check include:

- consult local staff to assess risk
- consult local police for advice, assessment and security.

5 Dredging techniques

Each type of dredging plant and method commonly used for inland dredging is described in this chapter. For each type, a general description is provided of the construction and method of operation. The advantages and disadvantages of each are described in relation to application, environmental effects and relative costs. However, it is important to understand that all comment is, of necessity, very generalised. Exceptions may arise in special situations whereby, despite being described here as generally poorly suited, a particular method may actually provide the optimum, or only, solution.

For each type of dredging plant, the environmental impact of the method is described in general terms and given an environmental impact 'rating' between 0 and 10. A higher rating indicates greater environmental impact if used in sensitive locations. The environmental rating is subjective, based upon the experience of the authors. The impact of each method is site dependent and full consideration of all aspects should be made when assessing the environmental impact of dredging at a particular site.

The characteristics of commonly used dredging plant are described in Table 3.1.

5.1 DRAGLINES

Draglines are almost exclusively operated from the land. Their main characteristics are illustrated in Figures 5.1, 5.2 and 5.3.

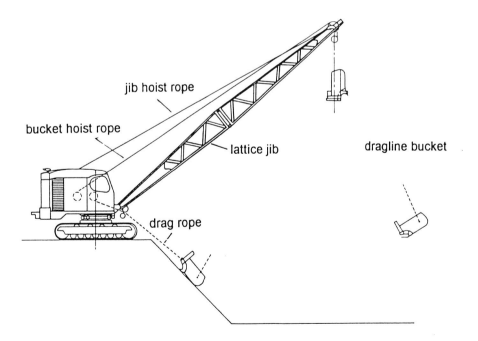

Figure 5.1 *Operation of dragline*

5.1.1 Mechanical characteristics

Dragline excavators usually comprise a track laying undercarriage on which is mounted a diesel driven system of winches. The machine module can be rotated through 360 degrees. A lattice crane jib, the angle of which can be adjusted, is attached to the machine module. The assembly is illustrated in Figure 5.1.

A toothed bucket, with open front and top, is suspended by wire rope from the extremity of the crane jib. A second wire rope is attached to the front of the bucket by means of a chain bridle. The two wire ropes are attached to hoist and drag winches respectively. The empty bucket is cast off, usually as the machine slews to the point of intended excavation. A skilled operator may cast the bucket over a distance from the machine up to 1.5 times the total jib length, though the level of control in such operation is imprecise.

The cast bucket falls to the bed and is then recovered by hauling on the drag rope whilst controlling bucket attitude and level by means of the hoist winch. When full, or reaching the nearest point of excavation, the loaded bucket is lifted and the machine swings to the point of discharge, when release of the drag rope causes the bucket to tip forward and discharge.

Figure 5.2 *Dragline excavator*

5.1.2 Operational effectiveness

The dragline is a simple machine and is best suited to simple tasks. Precise control is not possible, hence the machine is not suited to very accurate dredging, vertically or laterally. Due to its limited vertical accuracy the dragline is unsuited to maintenance dredging in clay lined canals. Cycle times (typically in the range of 50 to 55 seconds) are usually longer than for hydraulic machines of similar capacity.

Figure 5.3   *Dragline bucket*

Because attachment of the bucket to the machine structure is flexible, by wire rope only, the forces which can be applied to the ground to be excavated are limited, in the vertical plane to the self-weight of the bucket and in an approximately horizontal plane to the power of the drag winch. Hence the dragline may not be suited to capital dredging if the ground offers significant resistance to penetration by the bucket teeth.

Foreign matter, such as oil drums, large branches, etc., may be difficult to recover by dragline.

However, the machine should not be dismissed because of these short-comings. It also has advantages; those of simplicity and long out-reach relative to standard hydraulic machines. Where the required task is the removal of loose sediments from rivers, or ponds, in conditions of good and open bank access, and where the dredged arisings are to be spread over adjacent land, the dragline is a useful tool with modest operating costs. However, skilled operation is essential and the necessary skills may not be quickly acquired. Good operators are increasingly difficult to find.

5.1.3 Mechanical maintenance

Because construction is simple and robust, mechanical maintenance does not require a particularly high level of skill, or special facilities. Other than major engine repair, on site maintenance usually is possible.

Wire ropes require regular replacement and may be costly. Rope life is very much dependent on the nature of the dredging task and, more particularly, on the skill, care and routine maintenance provided by the machine operator.

Track wear may be expensive, particularly if operating in conditions of difficult ground, or where travel distances are long. Again, the level of care and routine maintenance by the machine operator is very important. Tracks may require regular adjustment to optimise efficiency and prolong life.

5.1.4 Environmental impact

Draglines have been used for many years in the maintenance and improvement of river channels. However, they are best suited to channels of regular section in open country. Modern trends, which favour irregular 'natural' channel sections, with the preservation or creation of pools, shoals and varied habitat, render traditional dragline methods less attractive.

In the area dredged, habitat destruction will be total, as with all dredging techniques. However, recovery may be quite rapid, particularly where dredging does not extend over the full width of the river bed or length of river bank (Pearson, 1975; Wade, 1978).

Turbidity levels will be increased significantly in the immediate locality of dredging. The rate of dispersion of the suspended sediments, and hence clearing of the water column, is dependent on the soil characteristics and water flow. However, generally the effects of increased turbidity and sediment dispersion will be localised.

The dragline requires more space in which to operate than other types of dredger. Hence it may not be practical to develop very varied bank side flora where regular maintenance by dragline is intended. Tree spacing must be wide and shrubbery must be limited to relatively isolated clumps. Reed beds and other vegetation adjoining the access bank may be difficult to preserve due to the method of operation of the dragline, wherein the loaded bucket is dragged to, and to some extent through, the water margin.

Environmental impact rating   5

5.2 GRAB DREDGERS

Grab dredging may be carried out from land or afloat. The grab bucket may be operated by wire ropes or hydraulically. Regardless of which system is employed, the general characteristics of grab dredgers are similar.

Rope operated grabs may be used by machines similar to the dragline. The only significant difference being in the arrangement of winches and wires.

Occasionally hydraulic grabs may be interchanged with the standard bucket of a hydraulic backhoe excavator (see Section 5.3). Details of rope operated grab buckets are shown in Figure 5.4 and a floating grab in Figure 5.5.

5.2.1 Mechanical characteristics

Whereas the dragline bucket is filled by dragging across the bed, the grab is lowered to the required level and loaded by forced closure of the opposing halves of the bucket.

If rope operated, the general structure of the machine is very similar to that of the dragline, indeed most draglines can be rigged for grab operation. If a grab machine is dedicated to dredging afloat, the tracked under-carriage may be omitted and machine module mounted directly on to a pontoon structure.

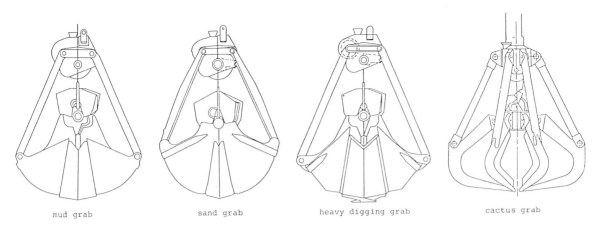

Figure 5.4 *Different types of grab bucket*

Figure 5.5 *Small floating grab dredger*

Grab buckets which are closed under hydraulic power are usually mounted on hydraulic machines, which may be alternatively configured as a backhoe (see Section 5.3).

5.2.2 Operational effectiveness

The grab dredger's method of working, whereby it dredges in a succession of spots, may be an advantage, or disadvantage. If the objective of the work is to dredge selected areas of very limited extent, such as in the formation of a trench, the use of a grab may be advantageous, but if the object is to clear a large area, such as a channel, to a specified level, then the grab is less well suited, due to the difficulty of achieving 100% bottom coverage. The result is usually a series holes and a very irregular bottom level. Substantial over-dredging is therefore necessary to ensure that no material remains above the specified level.

Cycle times of about 50 to 55 seconds for rope operated machines are generally similar to those of the dragline, but are longer than for hydraulic backhoes. Cycle times for hydraulic grabs may be slightly longer than for a similar machine operating in backhoe mode.

Subject to the character of the material to be dredged, control of depth is potentially superior to that of the dragline, but in reality considerable care is required to accurately control depth. The end result is often poor. Control of position is also potentially superior to that of the dragline, but is dependent on the sophistication of ancillary equipment fitted to the machine. If no equipment is fitted which indicates and records the position and effect of each dredging cycle, then control is dependent on the skill, concentration and memory of the operator. Generally, sophisticated equipment is not common on grab dredgers employed on inland waters. For this reason, the control and the quality of bottom finish usually is poor.

An important advantage of the grab is its insensitivity to debris and to oversize material. If the area to be dredged contains significant concentrations of debris, or awkwardly shaped oversize material, the grab dredger is likely to be the most effective dredging tool available.

5.2.3 Mechanical maintenance

For rope operated grabs, the level of maintenance is similar to that required for the dragline (see Section 5.1.3).

For hydraulically closed grabs, the overall level of maintenance required is broadly similar to that of the hydraulic backhoe (see Section 5.3.3). However, there are some significant differences. The grab bucket is more complicated than a backhoe bucket and hence requires more maintenance, particularly in respect of flexible hydraulic hoses, operating cylinders and fittings. In contrast, unlike the backhoe, it is less likely that heavy torsional, or shock loads, will be transferred from the bucket to the main machine structure. Hence the overall level of repair and maintenance may be slightly less for a hydraulic machine when operating with a grab bucket.

5.2.4 Environmental impact

The ability to dredge selectively may be an advantage when it is required to preserve selected areas of vegetation or bed. Some spillage of sediment from the grab bucket is unavoidable with conventional grabs, but where it is important to minimise spillage, special designs are available. These should be specified when a grab dredger is to be used for the dredging of contaminated sediments.

Design details range from simple plating of the upper bucket structure to minimise spillage due to over-filling, to special mechanisms which cause the bucket halves to close horizontally rather than radially.

However, it must be recognised that regardless of the type of grab bucket employed, spillage will occur if the bucket is not fully closed. Complete closure may be prevented by debris, or any substantial solid matter. In areas containing a significant quantity of debris, failure to fully close the bucket may be a frequent occurrence.

Because the grab bucket is delivered and recovered vertically, the method is more compatible with the use of silt screens than any other method. However, it should be noted that silt screens are unlikely to be very effective in moving water. The use of silt screens is discussed in Section 3.6.

Environmental impact rating 3

5.3 HYDRAULIC BACKHOES

Hydraulic backhoes may be pontoon-mounted for operation afloat, or on a wheeled or tracked undercarriage. The tracked undercarriage is the most common configuration (see Figures 2.1 and 5.6). Such plant dominates the maintenance of most inland waterways. This domination has been enhanced by the development of 'long-reach' machines, which are particularly well adapted to inland dredging. Long-reach hydraulic backhoes are available as a factory option from some machine manufactures (see Figure 5.7). Others are adaptations of standard machines modified by specialist companies. The development of long-reach hydraulic machines has contributed to the decline in the use of draglines.

Occasionally, hydraulic backhoes may be mounted on an amphibious structure. An example is the 'Watermaster' (see Figure 5.8), which also has an unusual option comprising a hydraulically driven centrifugal pump mounted within an excavating bucket.

In the UK, the main exceptions to the use of backhoe dredgers are situations such as wide rivers, lakes and ponds, where the extent of the area to be dredged is greater than the maximum outreach possible with even a long-reach hydraulic backhoe.

Figure 5.6 *Floating backhoe dredger*

5.3.1 Mechanical characteristics

All of the actions of the hydraulic backhoe are driven by linear or rotational hydraulics.

Excavation is achieved by a crescent-shaped bucket which can be rotated through approximately 150° in the vertical plane about a pin joint. Movement of the bucket is driven by a hydraulic cylinder via levers and links. The bucket is attached to the lower extremity of a rigid arm, usually known as the 'dipper', or 'stick'. The stick is attached to the extremity of a rigid 'boom'. Both stick and boom are of steel box girder construction. On long-reach machines the stick and boom are increased in length to provide a greater working radius. This results in reduced digging power, but this is not usually important in maintenance dredging. Connection of these components, and with the main machine structure, is by pin joints, which permit rotation in the vertical plane through an angle of between 150° and 180°. Movement is powered by high pressure hydraulic cylinders.

The overall action of the backhoe when excavating is analogous to the human hand and arm when scooping soil from the ground.

The main machine structure houses one, or in large machines, two diesel engines which drive hydraulic pumps. The operator's enclosure is offset at the front of the main machine structure. The complete machine assembly can be rotated through 360° relative to the base, which may incorporate wheels or tracks. Alternatively, machines which are dedicated to dredging afloat may be pedestal-mounted on a spudded pontoon (see Figure 5.9).

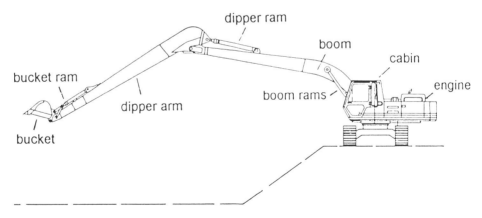

Figure 5.7 *Long-reach backhoe*

Variations on the fully hydraulic backhoe are available. The most common variation is the combined use of wire ropes and hydraulics in the digging action, as is provided on a range of machines knows as 'VC' machines (see Figure 5.10).

5.3.2 Operational effectiveness

The fact that the hydraulic backhoe has various advantages relative to other types of excavator is demonstrated by its rapid move to a dominant position on the earth-moving market.

Figure 5.8 *Amphibious dredger*

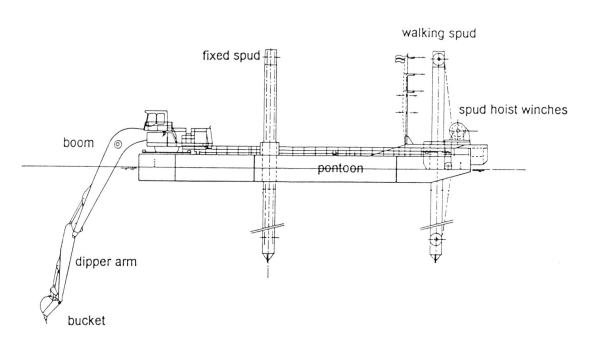

Figure 5.9 *Backhoe mounted on spudded pontoon*

Figure 5.10 *Long-reach machine with mixed hydraulic and rope operation*

The bucket can be placed with precision. The combination of rigid connection between the bucket and main structure, and hydraulic power, allows a wide range of ground types to be successfully excavated. These may range from mud to rock, though only weak rocks can be readily excavated, and these only by relatively large machines configured for operation at short radius. The backhoe is capable of restricted working in spaces which too narrow for a dragline or grab to work effectively, such as between bank-side trees or structures.

Awkward items, such as boulders, drums and tree trunks, can usually be handled by gripping the object between the bucket and dipper arm.

The action of the linear hydraulics is more rapid than winches and wire ropes and hence cycles times in the range 35 seconds to 45 seconds are faster than those of equivalent rope-operated machines.

A variety of different bucket types are available for different tasks. These range from very narrow buckets to form slit trenches to very wide buckets for ditch clearance. More complex attachments are available for tasks such as weed removal (see Figure 5.11) and sediment de-watering or screening (spinning bucket).

5.3.3 Mechanical maintenance

Hydraulic machines are significantly more complex than rope operated machines.

Routine maintenance which can be carried out on site includes greasing and oiling of tracks and slowing mechanisms. These should be carried out daily by the operator.

Figure 5.11 *Weed removal attachment (rake)*

Periodic maintenance includes oil changes, filter changes and replacement of worn or damaged parts, including hydraulic hoses and bucket teeth, and track adjustment. These may be carried out by the operator, or a visiting trained fitter, providing that drained oils and redundant filters are carefully collected and disposed.

All hydraulic power systems comprise moving parts which are manufactured to very small tolerances. If impurities enter the system, particularly abrasive matter such as sand, serious damage may result. It is also very important that the system, particularly the hydraulic pumps, is never allowed to run dry or experience seriously restricted oil flow. It is therefore essential that hydraulic oil level be maintained within the manufacturer's specified limits at all times.

5.3.4 Environmental impact

The potential impact of the backhoe is broadly similar to that of the dragline and grab, falling approximately mid-way between the two. The backhoe can be used to remove material more selectively than the dragline, but in most situations is likely to cause slightly more disturbance and suspension of sediments than an equivalent grab. If working within a silt curtain, the resulting restriction on efficient operation will be slightly greater than for a grab dredger.

Control of depth is superior to both dragline and grab, particularly if dredging in medium or stiff clays. This allows better control of the finished bed level with greater scope for achieving a particular profile and for minimising unwanted over-dredging. Long-reach machines can sometimes dredge from a field or an access track which is set back from sensitive waterside margins, thus causing less disturbance.

The failure of high pressure hydraulic pipes is not uncommon and non-toxic vegetable or synthetic oils should be used in all hydraulic machines when working in an aquatic environment.

Environmental impact rating   4

5.4 PLOUGHS, RAKES AND WATER INJECTION

None of these can be properly described as dredgers. They do not remove material from the system, but merely move it within the system. During the 1970s and 1980s the plough became very popular within the port and harbour dredging industry. Its potential within the inland waterway system is less, but both the plough and rake may be useful tools under appropriate circumstances. Where conditions are appropriate, bed material may be moved at lower cost by the use of a plough than by any other method.

Ploughs may be used to drag material over short distances, typically about 50 m, from shallow areas into deeper areas, i.e. from shoals to scour holes.

Rakes can be used to force sediments into suspension. The suspended sediments may then be transported from the work area by the prevailing current. This is termed 'agitation dredging'. However, it is inevitable that a short-term reduction in water quality will occur and results are unpredictable. Rakes are not recommended for the bulk movement of bed material.

Rakes may be used to remove debris, or over-size material from the bed prior to dredging (see Figure 5.12). This may be necessary if suction dredging methods are to be employed.

Figure 5.12 *Use of a backhoe fitted with a rake to remove debris*

Both ploughs and rakes are towed at a controlled height relative to water or bed level behind a suitably equipped work-boat or tug. A typical work-boat and plough are illustrated by Figure 5.13.

For ploughs or rakes to be effective, certain conditions must prevail:

- minimum water depth throughout the proposed working area must be greater than the work-boat maximum draught; maximum draught may increase under towing loads
- water width in the working area, or close by, must be sufficient for the boat to be turned
- areas from which sediment is to be removed must be close to areas deep enough to accommodate the material moved
- the sediments must be loose and preferably fine grained
- if sediments are to be transported in suspension, water currents must be sufficient for the purpose
- if sediments are to be transported in suspension, no harm should result at the place of final deposition.

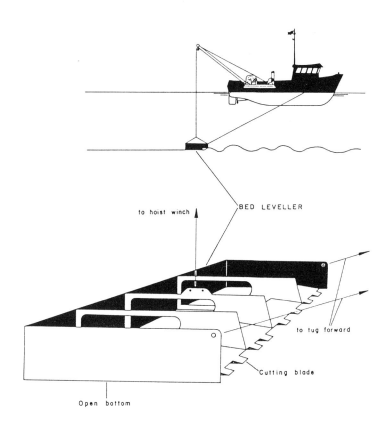

Figure 5.13 *Workboat and plough*

Water injection methods move bed material by reducing its *in situ* density to the point when it will flow as a dense fluid under the influence of gravity. The change in density is achieved by injecting water into the bed from a series of low pressure, high volume jets. Water is supplied by pump to the jet nozzles which are closely spaced along a horizontal T-bar construction. The height of the T-bar jet head relative to the bed is controlled by a winch or hydraulic cylinders. The principle features are illustrated in Figure 5.14.

Subject to site conditions, relatively simple water injection methods are capable of achieving high rates of production, moving sediment over considerable distances. In appropriate circumstances the unit cost of moving sediment may be low relative to other methods. If the method is properly employed, there should not be a significant increase in the level of suspended solids in the upper water column. However, the operation can not be fully controlled and hence the fate of material moved may be uncertain. <u>Experienced assessment of the likely end result is necessary before water injection methods are employed.</u>

The conditions which must be satisfied if water injection methods are to succeed are broadly similar to those applicable to ploughs and rakes, but with minor variations, as detailed in the following list:

- minimum water depth throughout the proposed working area must be greater than the maximum draught of the tug-boat and the full extension of the water injection equipment
- the sediments must be loose and preferably fine grained
- substantial transport will only occur for fine grained particles (silt content)
- the presence of water currents is not a requirement
- the assessed final place of sediment deposition (i.e. an area adjacent to the dredging site) must be able to accommodate the volume of material moved and the deposition should not cause harm.

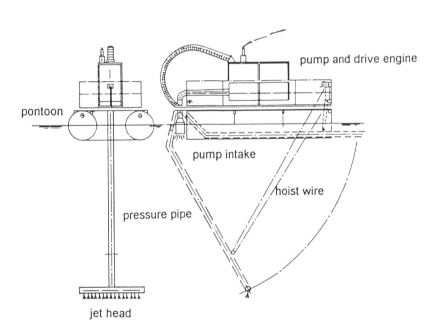

Figure 5.14 *Simple water injection equipment*

5.4.1 Environmental impact

If a plough is used to move bed material from shoal areas into deeper areas close by, the resulting increase in turbidity due to the suspension of fines is likely to be slightly greater than if a dragline is used. This is because scour by the wash from the propeller of the towing vessel is likely to force part of the loosened sediment into suspension. The effect will be less in sandy bed material, or in greater water depth. The effect may be negligible where water depth exceeds the stern draught of the towing boat by more than 3 m.

If the bottom sediments are fine grained, that is less than 0.06 mm, and current velocity exceeds about 0.3 m/sec, ploughing may cause a significant volume of sediment to be suspended and this may be transported over considerable distances. A very rough approximation of transport distance may be made by determination of the particle settling velocity using Stokes Law (see Appendix). This applies to fine sand grade material and presumes that once the particle has fallen to the bed it will remain stationary. This will only be true if the current velocity is lower than that necessary to produce critical shear stress at the bed.

The potential effect of the redistribution of fine sediments over large areas downstream must receive careful consideration before agitation or water injection methods are employed. If the downstream habitat is sensitive to raised levels of turbidity, or fine grained blanketing of the bed, such methods should not be used. Nor should the method be used where the downstream water, or downhill bed, is in the ownership of another party, or occupied by sensitive commercial processes, such as a fish hatchery or paper manufacturer.

Environmental impact rating - plough   6

Environmental impact rating - rake   7

Environmental impact rating - water injection 6

5.5 BUCKET DREDGERS

Bucket dredgers and cutter suction dredgers (see Section 5.6) are designed only for operation afloat. Each is described in greater detail by BS 6349 Part 5: 1991.

The bucket dredger, sometimes called 'bucket chain dredger', or 'bucket ladder dredger', has been used for many years in the improvement and maintenance of inland waterways. However, it is suitable only for waterways where section and water depth are adequate. The use of bucket dredgers in rivers is unusual, except occasionally in tidal reaches or on larger rivers such as the Thames, because they are not suited to operations in very shallow water. The main features of the bucket dredger are illustrated by Figure 5.15.

Their construction is usually on a 'one-off' basis and several crew are required for efficient operation. These factors result in higher capital and operational costs than those for draglines, grabs and backhoes. For this reason, popularity has steadily declined over the past 3 decades, particularly in the UK. However, their use remains relatively popular on the inland waterways of Belgium, France, Germany and Holland. The reason for this is not clear, but probably stems from the greater size and extent of inland waterways in mainland Europe. Under suitable conditions, the bucket dredger is potentially more productive when dredging inland waterways than any other type of dredger. This is because the dredging process is continuous, whereas with all other types, other than the cutter-suction dredger (Section 5.6), production is intermittent.

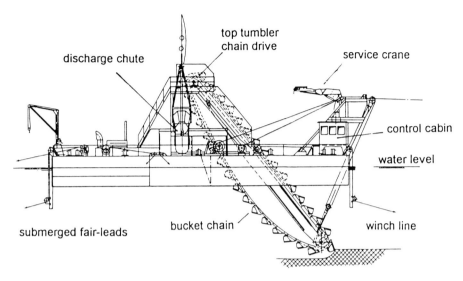

Figure 5.15 *Schematic of a bucket dredger*

5.5.1 Mechanical characteristics

The bucket dredger comprises a continuous chain of linked buckets which excavate and convey soil. Individual bucket capacity may range from 150 litres to 1,100 litres. Only smaller capacities are commonly used on inland waterways in the UK.

The steel bucket chain, which is of heavy construction, is supported by a large beam of box girder construction termed the 'ladder'. The upper end of the ladder is supported by a hinge mounted well above deck level. The lower end is supported by a hoist winch which controls the ladder position and depth of cut. The bucket chain passes over, and is driven by, the 'top tumbler' which is usually powered by a hydraulic or electric motor. On old machines the top tumbler may be belt driven from an engine on deck. The lower end of the chain passes round a free tumbler.

The dredging action is achieved by bed material being scooped up by buckets moving round the lower tumbler. The loaded buckets travel up the inclined ladder, and are inverted and discharged on passing over the top tumbler. The discharged material falls onto a steeply sloping chute which conveys it to a hopper barge moored alongside. Very occasionally machines are adapted to discharge to the shore via a conveyor or pump.

The whole machine is advanced into the material to be dredged by pulling on a head winch. Lateral movement is achieved by side winches (see Figure 5.16). Winch wires may be attached to anchors in wide expanses of water, or to suitable fixed points ashore when working in narrow waterways.

5.5.2 Operational effectiveness

Because the dredging action is continuous, production rates may be high relative to other types of dredger. The pattern of working, which involves an advancing 'sweeping' action (see Figure 5.16), allows a level finish to be achieved with good vertical accuracy. The digging action is quite powerful, which allows a wide range of material types to be dredged effectively. The open bucket structure permits quite a wide range of debris to be dredged with only modest disruption of the process.

The 'multi-winch' method of location and movement may cause an obstruction to passing traffic, though this problem may be alleviated by routing anchor wires via underwater fairleads. It may sometimes be difficult to locate suitable anchorage points required by the multi-winch system.

With rare exceptions, bucket dredgers can only be used to load floating hopper barges and can only work with a barge moored alongside. These place limitations on the method of disposal and the minimum width of waterway in which operation is possible.

Their construction is usually heavy and monolithic. Hence movement from site to site may only be possible by water.

5.5.3 Mechanical maintenance

Whilst the overall construction of the bucket dredger is simple and rugged, there are many areas of steel-to-steel contact between moving parts. Wear rates may sometimes be high. The unique design and construction may result in spare parts being expensive relative to standard production machines, such as small to medium hydraulic backhoes.

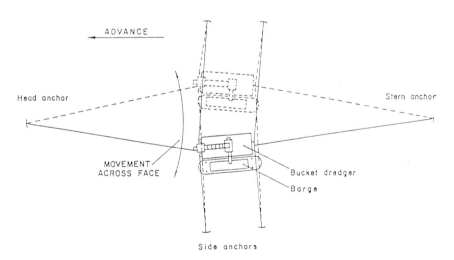

Figure 5.16 *Mooring pattern and method of operation of a bucket dredger*

Routine maintenance such as oiling and greasing is important if wear is to be minimised, but is easily achieved by non-skilled labour on site. The power requirements of bucket dredgers are modest and hence diesel engines need be of only modest capacity and simple construction. Hydraulic drives to winches are also simple and need not operate at high pressure.

The main areas of difficulty in the maintenance of bucket dredgers stem from their heavy construction, which at times will require the use of heavy lifting equipment, particularly if the bucket chain must be removed or adjusted. Maintenance of the hull and other components below the water line will necessitate removal from the water using a slipway, or a dry dock, which may be expensive.

Maintenance costs will be minimised by a diligent and well-trained crew. The life of wire ropes in particular is strongly influenced by the level of care and maintenance provided by the crew.

5.5.4 Environmental impact

The bucket dredger is not well suited to selective dredging, being most effective if dredging to a fixed level and width over substantial channel lengths. Increased water turbidity will be caused by spillage, or washing, from over-filled buckets, or material which fails to fall from the buckets on initial inversion, but is washed off during descent through the water column. Spillage may also be caused by splashing during discharge to the transport hopper barge. All of these factors can be minimised by careful operation, particularly by careful control of the rate of advance.

If operated conscientiously by a skilled crew, the volume of spillage can be restricted to levels similar to, or lower than backhoes and grabs. However, the measures necessary to minimise spillage will also result in lower rates of production.

If a bucket dredger is to be used to dredge contaminated materials and no significant spillage is acceptable, there are various modifications which may be made to the dredger which will permit dredging with minimal spillage (Anon., 1994).

The specific energy consumed, that is the amount of energy expended in dredging and discharging one cubic metre of soil, is lower than for any other type of dredger.

Environment impact rating - standard dredger   3

Environment impact rating - modified dredger   2

5.6 CUTTER-SUCTION DREDGERS

With a few minor exceptions, cutter-suction dredging is conducted afloat. Dredged material is discharged by pumping through a pipeline. The dredger is not well suited to the loading of hopper barges, unless the material dredged is coarse grained, such as sand and gravel. The ability to dredge and transport bed materials in a single independent closed process is unique amongst inland dredging plant. The main features of the cutter-suction dredger are illustrated in Figure 5.17.

5.6.1 Mechanical characteristics

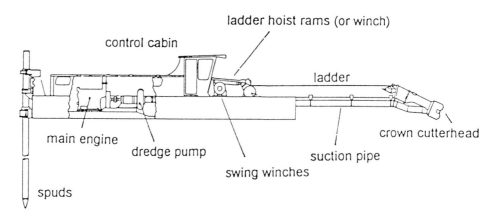

Figure 5.17 *Schematic of a cutter-suction dredger*

The cutter-suction dredger is quite unlike any other type of dredger engaged on inland waterways. Other types all achieve initial excavation and movement of sediment by various types of bucket. The cutter suction dredger moves material by pumping. Initial dislodgement of the bed material may be achieved by one of a variety of mechanical cutters. The cutter breaks out material from the bed and directs the free material towards the pump suction intake, which is located close to, or within, the cutterhead. A cutterhead is necessary to achieve maximum production and control of the dredging process, even in loose materials. The soil and water mixture is passed through a pump (usually centrifugal, but very occasionally piston) to a floating discharge pipeline connected to the stern of the dredger. The floating pipeline connects to a pipeline on shore, through which the mixture is pumped to a purpose-built embanked containment area.

The suction pipe and cutterhead are supported by a fabricated steel ladder which is connected to the main hull by hinged pin joints. The free end of the ladder is supported by a hoist winch, or hydraulic cylinders, by which means the depth of dredging is controlled. The main pontoon usually contains the engines, pump, hydraulics, electrics and operator controls.

Movement of the dredger and the pattern of movement are dependent on the type of cutterhead employed.

If a 'crown' type cutter is used (see Figure 5.18), movement by a combination of forward winches and stern spuds will usually be most effective. Stern spuds may be fixed or one may be mounted in a moveable carriage. Movement is by means of a hydraulic cylinder, usually with a maximum stoke of between 4 m and 6 m. Winch wires are attached to anchor points located alongside the dredger. The dredger works by swinging from side to side about a single stern spud. Multi-winch systems may be used without spuds, but the level of control of position is inferior.

If an 'auger' type cutter is employed (see Figure 5.18), the dredger is usually advanced longitudinally in a series of parallel cuts by a single winch. The winch cable is tensioned between anchor points located forward and aft of the dredger. Alternatively, larger or more sophisticated auger dredgers may be advanced by a spud carriage.

Winches are most commonly driven by hydraulic motors, but occasionally electric drives may be used.

In addition to the common types of cutter-suction dredger described, there are a wide variety of special designs, including some which are amphibious.

5.6.2 Operational effectiveness

The cutter-suction dredger shares with the bucket dredger the characteristics of sweeping coverage of the area dredged and a continuous dredging process. A uniform level bottom finish and relatively high rates of production are therefore possible. However, unlike the bucket dredger, if the common centrifugal pump is used, the material dredged is highly diluted by mixing with water to facilitate hydro-transport.

Figure 5.18 *Crown and auger type cutter heads*

The percentage of soil in the total mixture may range from 0% to 30%, but is unlikely to average more than 20%, and most commonly the average will be in the range 10 to 15% (see Section 5.10). These percentages are based on the ratio of the volume of soil contained in the pumped mixture, at its *in situ* density, and the total pumped mixture volume. At a given concentration by volume, the density of the pumped mixture is dependent on the *in situ* density of the material to be dredged. Concentration may also be expressed as the percentage of the weight of dry solids in the total mixture, but whilst this relationship is more convenient for calculation, it has less relevance to the real situation.

A few cutter-suction dredgers employ a fixed displacement piston pump. Such systems are not yet fully developed, but hold the promise of pumping solids at much higher concentrations than is possible using a centrifugal pump. Soft muds can be pumped without dilution. This may provide important practical and economic advantages if water supply is limited, if excess water complicates transport, or disposal, or when the materials to be dredged are contaminated.

Cutter-suction dredgers are constructed in a wide range of sizes and installed power. However, those used for the dredging of waterways in the UK are usually small, with discharge pipe diameters in the range 150 mm to 300 mm. The maximum size of solid sphere which can pass through a centrifugal pump is approximately 80% of the discharge diameter. Thus, the size of solids, such as cobbles or boulders, which can be handled by pumps in common use is in the range 120 mm to 240 mm, and these only as an isolated occurrence. Significant concentrations of material of these sizes will cause blockage. The time required to clear a blockage may be substantial. For this reason, the cutter-suction dredger is not well suited to the dredging of areas which contain significant concentrations of debris, or other over-size solid matter. Dense, or fibrous vegetation may also disrupt dredging and should therefore be removed in advance by other methods. Some types of vegetation may be undermined by dredging, and when floating freely, removed by other methods.

Dredgers fitted with crown cutterheads are able to dredge a wide variety of soil types ranging from mud to stiff clay, or even very weak rock, such as soft chalk. Large cutter-suction dredgers are able to dredge a range of weak or fragmented rocks.

Because cutter suction dredgers discharge dredged material via a pipeline as a dilute slurry mixture, the area into which the slurry discharges generally must be fully enclosed. Enclosure is usually achieved by the construction of earth banks around the perimeter of the land area to be used for disposal (see Section 5.9.4). It is also possible to discharge to hopper barges, or road tankers, but this is rarely economic due to the difficulty of separating the soil from the added transport water. However, such methods may be worthy of consideration in special circumstances. For example, clean medium or coarse sand will settle from suspension quickly and it is then possible to load hopper barges satisfactorily.

Occasionally discharge may be by spraying the pumped mixture through a nozzle fitted at the stern or side of the dredger, or at the end of a very short length of pipeline. The method is known as 'rain-bowing'. The initial result is invariably very messy, particularly in windy conditions. The entire bank-side area of discharge is covered in a thin layer of wet soil. However, if the bank-side land is agricultural and the dredged soil of a type which will bring nutritious or structural improvement to the land, rain-bowing may be appropriate. The distance over which material can be sprayed is dependent on the characteristics and power of the dredge pump, but for a small dredger will typically be in the range 10 to 25 m.

5.6.3 Mechanical maintenance

Centrifugal pumps impose high and fluctuating loads on the power source. Engine life may therefore be less than for other applications. An engine with good torque characteristics is most appropriate.

Hydraulic systems usually operate at quite high pressures, typically about 150 bar. The system is therefore highly stressed and pipe failures are not uncommon. Cleanliness is important, but not easily maintained aboard small floating dredgers. Within the engine and pump room, which on small dredgers are usually one room, the combination of a hot diesel engine and the large cold mass of the dredge pump may result in heavy condensation. This may cause more rapid corrosion of metals and deterioration of the electrical installation than is usual.

Regular oiling, greasing and cleaning is essential. Oil levels in engines and hydraulics should be checked at least on a daily basis and preferably at shorter intervals. Regular re-painting of metalwork is necessary to prolong life. At all times wire ropes should be properly laid on drums and damaged ropes replaced.

5.6.4 Environmental impact

If operated with care, the level of spillage and suspension should be very low. Auger cutterheads are more effective in minimising spillage and suspension than crown types. Once the dredged mixture has entered the suction intake it should be totally contained until discharged at the disposal site. Interaction with the surrounding aquatic and bank-side environment is therefore eliminated. This may be an important advantage if dredging contaminated material.

The level of spillage and suspension caused by appropriate and well operated cutter-suction dredgers is lower than is possible with any other type of dredger.

All material is transported by pipeline. Hence, following installation of the discharge pipeline, no bank-side traffic is necessary. This may be an important advantage if the waterway passes through environmentally sensitive areas, high quality farmland in productive use, or through congested, or developed areas. For this reason, pumping methods are commonly used for the de-silting of ornamental lakes.

The specific energy consumption of a conventional cutter-suction dredger is quite high, due to the large volume of water which must be transported with the soil. On most sites, it will be possible to return draining water from the disposal site to the area of dredging by gravity, but occasionally water must be returned by pumping, in which case energy consumption is increased further.

Apart from relatively high specific energy consumption, the cutter-suction dredger may be the most appropriate dredging tool where protection of the aquatic or bank-side environment is paramount. However, the method can only be used economically when the area to be dredged is free of significant concentrations of debris, or oversize material, and where a suitable disposal site is available within a reasonable pumping distance. Pumping range will be dependent on the power available, the pump characteristics, the sediment characteristics and the elevation of the disposal site relative to the site of dredging (see Section 5.9.4).

Dilution of the soil during pumping results in a large temporary increase in volume. The capacity of the contained area into which the mixture is pumped must be sufficient to accommodate this increase in bulk.

Environmental impact rating   *1*

5.7 WINCH ENGINES

The use of winch engines, sometimes called 'slack lines', is not common, but may occasionally be found and may be suited to special situations. The following brief description outlines the principles.

5.7.1 Mechanical characteristics

The name 'winch engine' dates back to the use of steam powered traction engines with a rear mounted winch. These were commonly used for ploughing fields.

In dredging applications, either two engines, or one engine and an anchored return pulley, were used. One engine was placed on the bank of the lake, or area of water to be dredged, and the other engine, or pulley block, on the other. Wire cables from each side were attached to the front and rear of a large scraper bucket, similar in design to a dragline bucket (Figure 5.3) but usually much wider. The bucket was pulled underwater across the bed gradually filling as it progressed. On reaching the near-side bank the bucket was hauled up the bank and over a temporary ridge or ramp to discharge under gravity. The bucket was then hauled empty to the far side water edge and the cycle repeated.

The method can occasionally still be found in use today, though the use of steam engines is rare, being mostly confined to steam enthusiasts. For a more commercial operation, steam engines can be replaced by large truck-mounted hydraulic winches, as seen on large recovery vehicles.

5.7.2 Operational effectiveness

The obvious advantage of the system is that it is land based, but in contrast to all other land-based methods, has almost unlimited out-reach. Control of the depth of dredging and thickness of sediment removed is limited. If the area to be dredged comprises soft silts overlying firm ground, the bucket will tend to ride over the firm ground and remove only soft superficials. Some degree of control of penetration into medium dense bed material is possible by adjustment of the bucket cutting blade and hauling bridle attachment.

Whether employing traditional steam engines, or modern truck mounted winches, the engine-winch unit is heavy and not well suited to operation on soft ground.

5.7.3 Mechanical maintenance

This is a simple system, comparable to a dragline, but on larger scale. Modern winches are usually mounted on a wheeled chassis and hence maintenance is straightforward. Care is necessary to avoid the unit sinking in soft ground, and to maintain the pulling cables in good order by careful spooling on the winch drum.

5.7.4 Environmental impact

The method has a destructive potential which is probably unequalled amongst dredging plant. All vegetation and immobile life on the bed is likely to be destroyed. Bank vegetation on the bank on which the winch engine is placed will also be destroyed. On the plus side, there is little disturbance to the far bank, especially if a return pulley block system is used. Agitation and suspension is relatively modest, though sediments and water spilling from the bucket during recovery and hauling up the home bank will cause localised turbidity along the water margin.

Environmental impact rating   10

5.8 LOW TURBIDITY DREDGING

Increased water turbidity will result if the dredging process releases sediment into the water at a significant rate. Usually it will be desirable to minimise any increase in suspended solids, but particularly so if the local aquatic environment is sensitive, or if the dredged material is contaminated.

Regardless of the method of dredging employed, the release of soil will be controlled by the partial enclosure of the dredging process. This may be achieved at source, by enclosing the point of excavation, or remotely, by enclosing or isolating the area of dredging operations. The latter may be partially achieved by the use of silt curtains (see Section 3.6), or in the case of canals, the area might be fully isolated by damming at the limits of the area to be dredged.

Total enclosure at the point of excavation, that is at the cutting edge of the bucket or cutter head, is clearly not possible. However, varying degrees of shielding are possible. This is most easily achieved when using suction dredging methods in conjunction with special cutterheads, such as the horizontal auger type cutterhead. In the latter case, the top and ends of the cutter can be covered by fixed or adjustable metal shields. Various designs of cutterhead and enclosure have been developed specifically to minimise the escape of soil particles during dredging.

Bucket dredgers may be modified by enclosing the bucket chain within a tunnel, fitting air release valves to the buckets and washing jets to clean the buckets after discharge.

The bucket of grab dredgers may be plated so that material is not spilled due to overfilling. Special grab buckets are available which close horizontally rather than radially. This permits greater control of the excavating process, particularly when a relatively thin surface layer of sediment is to be removed.

Backhoe dredgers are not easily improved due to the nature of their excavating method.

Where it is important to minimise the release of fines one of the methods described above should be adopted. That which is most appropriate is dependent on the specific site conditions and nature of the work to be achieved. Generally, suction methods are most easily adapted to minimise turbidity because each of the processes of excavation, discharge and transport can be almost totally enclosed. However, suction methods are well suited only to the dredging of material which is relatively free of debris, cobbles and boulders. Contaminated soils are most commonly found in urban or industrial areas where the concentration of debris may be high.

5.9 SEDIMENT TRANSPORT METHODS AND EQUIPMENT

The choice of transport method may be dictated by either the method of dredging or place of disposal. However, because the transport method may have important environmental and economic effects, it should be considered alongside other matters early in the project assessment.

Environment is considered only generally in the description of each transport method, but relative environmental impact factors are provided to compliment those for the primary dredging methods.

5.9.1 Overland

Overland means cross country, as opposed to over metalled roads. Overland transport may be by dumpers or trucks, but most commonly will be by dumpers because of their ability to negotiate a wider range of ground conditions.

All overland transport will cause damage to the land. This may be due to compaction, rutting and grass or crop destruction. Such damage may be reduced during summer, or dry periods, but the different problem of dust may arise. Damage may be eliminated, or minimised, by the construction of a temporary haul road using geotextiles and imported granular fill.

The specific energy used in overland transport is higher than for all other transport methods. This is due to the relatively high power of all terrain vehicles, the higher rolling resistance relative to metalled surfaces and the fact that a redundant return journey is necessary. Supplementary plant, such as a grader or bulldozer, may be necessary for haul road maintenance.

From an environmental viewpoint, overland transport should be considered one of the least friendly methods. However, the same may not be true when spreading dredged material over land. Spreading may be achieved without the concentration of vehicular traffic along pre-defined routes (see Section 2.7).

Environmental impact rating   7

5.9.2 Road

Specially constructed vehicles are necessary for the transport of wet soil over public roads so that spillage does not occur.

It is not common for the site of dredging work to adjoin a public road, hence some element of overland transport to gain access to the road system is likely to be necessary. This inevitably will result in some fouling of the highway by mud or dust. Subject to the particular site conditions, it may be necessary to provide vehicle washing, or dust suppression facilities, prior to access to a public road.

Road congestion in many areas is already severe and the transport of dredged material by road will increase any existing congestion. The trend towards larger and heavier trucks is the cause of an increasing rate of damage to the infrastructure. Noise levels are increased by large trucks. The use of large trucks within the community generally is resented by the local population. Overall, the adverse effects of haulage of dredged material over public roads is serious and the method must be considered to be the least attractive of the available methods. This is reflected in the award of a high environmental impact rating.

Private road haulage companies are numerous and hence competition is usually strong, resulting in highly competitive pricing. This may distort the cost of road transport relative to other methods, but where the real cost provides economic advantage, this should be considered together with other matters.

Environmental impact rating   8

5.9.3 Hopper barge

The use of hopper barges has the advantages of minimal environmental impact and little or no interaction with agriculture or the general community, other than navigators, who usually will be the beneficiaries of dredging work. The specific energy consumption in transport by hopper barge is usually less than any other method. It follows that transport by barge should be considered favourably in all situations where their use is feasible. That is, in all situations where the waterway between the sites of dredging and disposal is navigable. Even when only part of the route is navigable, barge transport should be considered if its use significantly reduces reliance on overland or road transport.

For efficient operation the output of the dredger must be equalled by the total transport capacity of the barges. If the dredger is to work continuously, which is usually desirable, more than one barge is required. The total number of barges required is dependent on the barge cycle time, that is, the total time from completion of loading to arrival back at the dredger after transit and discharge. Transit time is dependent on the total haul distance, speed of travel and time to negotiate locks or obstructions, if any. The number of barges needed for continuous production is then found by dividing the barge cycle time by the time taken to load one barge, and adding one. It is important to correctly determine the number of barges necessary for efficient and economic operation before work commences. It is often necessary to strike a balance between limited idle time for the dredger and the cost of employing an additional barge.

Hopper barges of course should be maintained in good condition, free of leakage, and should not be over-filled so that spillage occurs. The size of barge should be appropriate for the site conditions. Oversize barges which cause a high blockage factor should be avoided, as should excessive towing speeds. The blockage factor is the relationship between the cross-sectional area of the submerged hull of the hopper and that of the waterway. If the barge size is large relative to the waterway, waves and strong currents will be generated causing erosion of the banks. It will be prudent to apply conservative towing speeds during project evaluation.

Most hopper barges used on inland waterways in the UK are dumb, that is, without independent propulsion. Movement is by tug, which should also be well suited to the site conditions. Barges may be moved by towing or pushing. Underkeel clearance should be sufficient to minimise the risk of bed erosion, or propeller fouling by bottom debris.

It is uncommon for hopper barges to be fitted with equipment to facilitate self-discharge. Discharge is usually by means of a land-based grab or backhoe (see Figure 5.19). However, at sites which experience large or regular dredging operations, barges may be discharged by pumping. The pumping installation may be permanent or mobile and, in order to discharge, the dredged material is fluidised by a high pressure water jet and pumped from the hopper in suspension. The pumped mixture is conveyed by pipeline to a purpose-built disposal lagoon (see Section 5.9.4 and Appendix.

Environmental impact rating   1

Figure 5.19   *Discharge of hopper barge by land based backhoe*

5.9.4 Pipeline

Transport by pipeline (see Figure 5.20) is usually only associated with dredging by cutter-suction dredger (see Section 5.6). A notable exception is the Watermaster dredger which combines the characteristics of the hydraulic backhoe and a submersible centrifugal pump. However, power supply to the pump is modest and hence the maximum pumping distance is also modest relative to a conventional cutter-suction dredger.

Pump discharge can be adapted for use with other methods, including bucket and grab dredgers. However, the combined dredging and pumping processes are relatively complicated and are not common. Nevertheless, if the circumstances of a particular project favour transport by pipeline, but not the use of a cutter-suction dredger, the combination of mechanical dredging and hydro-transport may be worthy of investigation.

Figure 5.20 *Transport by pipeline*

The general practitioner engineer will usually be familiar with transport by dumper, or truck and transport by hopper barge is straightforward. In contrast, transport by pumping through pipelines is less routine and some simple guidelines are therefore provided in Appendix.

Containment of pumped material

It will usually be necessary to contain pumped material to restrict its spread and to impound water to achieve conditions conducive to the settlement of soil grains.

Containment is usually achieved by the prior construction of earth bunds to form one or more enclosures (see Figures 5.21 and 5.22). If the dredged material is fine grained, or organic, the total volume of the containment area must be substantially greater than the volume occupied by the material *in situ* prior to dredging. Internal division of the area may be necessary to permit filling on a rotational basis, thus allowing material to be deposited in thin layers, to drain and to consolidate, prior to further filling.

Following dredging, pumping and deposition, the bulk of most soil types will increase. The capacity of the containment area must be sufficient to accommodate the increased bulk. Further capacity is necessary to allow a minimum depth of impounded water at all stages of filling. It is desirable that the minimum depth of supernatant water (water above the level of settled soils) be 200 mm. Greater depth may be necessary for fine grained or organic materials because their rate of settling is much slower. Bund height should be such that the minimum depth of supernatant water can be maintained during the final stage of filling. A rough guide to the additional capacity which should be provided for various types of dredged material is given in Table 5.1.

Figure 5.21   *Suction dredger pumping to remote containment areas*

Table 5.1   *Guide to contaminant area total capacity*

Material type	Capacity factor	Minimum depth of supernatant water
Organic	2.0	300
Soft clay and silt	1.6	250
Sandy silts	1.5	200
Fine to medium sand	1.3	150
Gravel	1.2	100

Note: To find required containment capacity, multiply *in situ* volume to be dredged by capacity factor.

Figure 5.22 *Extensive containment areas*

Additional bund height will be required to provide an adequate freeboard. The necessary freeboard is dependent on the type of soil used in bund construction, bund width, the method of construction and the importance of achieving 100% security.

A more accurate assessment should be made during project planning by means of a simple sedimentation test, wherein a sample of the soil to be dredged is mixed with water in similar proportions to those expected during pumping. The mixture should be poured into a graduated glass cylinder, thoroughly agitated and allowed to settle. The height of the settled sediments should be recorded with respect to time. The initial settlement from suspension may range from a few seconds for sands, to perhaps one hour for fine grained materials. Consolidation of fine sediments will continue for much longer and the settlement records should therefore continue for several days. The resulting graph will provide useful guidance concerning the maximum capacity which is likely to be necessary.

A theoretical estimate of the settlement rate of fine soil grains may be made using Stokes Law (see Appendix).

Drainage

Water should be drained from the containment area at a level such that water is impounded to a depth above the level of settled, or settling, solids as indicated by Table A.5 in Appendix. Effective drainage will be best achieved using an adjustable weir, the length of which should be sufficient to restrict the depth and velocity of overflow water such that the entrainment of soil particles is negligible (Figure 5.23).

To allow the maximum opportunity for soil solids to settle from suspension, the drainage of supernatant water should be at a location as far as possible from the point of discharge of the pumped mixture. It may be advantageous to increase the total distance between discharge and drainage locations by means of internal bunds or training works. The minimum distance can be estimated using Table A.6 in the Appendix and knowledge of the rate of pump discharge and the containment area dimensions. However, the results must be treated with caution, because preferred drainage routes will tend to develop within the containment area. Current velocities within these routes will be much higher than across the section as a whole.

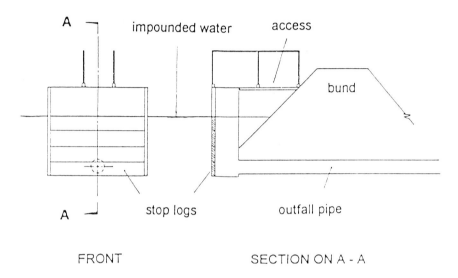

Figure 5.23   *Weir Box*

Environmental impact

The environmental impact of transport by pipeline should be minimal. The only exception, which should not arise, is in the event of pipe failure and consequent spillage of the pumped mixture. However, such events can occur and appropriate contingency plans should be made.

The only interaction between pipeline transport and the local environment occurs during the process of pipe laying. In most cases of inland dredging appropriate pipes can be laid manually. However, if machines are used, some minor damage to the bank-side or overland environment may be inevitable. Some clearance of scrub, or heavy vegetation may be necessary to facilitate pipe laying. If laid by hand, a 1 m corridor should suffice.

When the pipe has been laid, other than regular inspection and maintenance, no further access should be necessary until the pipeline is dismantled. In the interim all transport of the soil and water mixture is confined and isolated from the environment and general public.

Environmental impact rating   1

5.10 SOIL MODIFICATION

The structure of soil after dredging is affected by the dredging method. Soil structure may be important for various reasons. Soil structure affects the following:

- handling
- transport
- disposal
- beneficial use
- workability.

The main influence which the dredging method may have on the dredged soil is to change its water content.

If the dredged material is sand, drainage may be rapid following deposition. The drained deposit is then easily re-handled and, if necessary, easily transported. If the dredging is for maintenance, it is more likely that the soil will be fine grained with poor drainage characteristics. The time required for the release of excess moisture will then be prolonged, unless special measures are employed. Soils with very high moisture content are not easily handled, or reworked, are more difficult to transport and may be unacceptable at licensed disposal sites. These difficulties are compounded by the added weight of water which may further increase the costs of transport and disposal.

Suction methods of dredging, which are usually accompanied by pipeline transport, most commonly employ one or more centrifugal pumps. Pumping is then only possible following severe dilution of the bed material. The volume of *in situ* material contained within the pumped mixture may be a small percentage of the total mixture volume. A typical range of concentration is 5% to 35%, with 20% to 25% being a good average. Higher concentrations are possible if the material to be dredged is fine grained and low strength and if the excavating process is well controlled, but concentration levels above 30% are unlikely to be sustained due to interruption of the excavation process at the limits of excavation and when moving the dredger. Hence, unless special techniques are employed, the overall average concentration is unlikely to better 20% to 25% and may be much lower.

The added difficulties caused by high water content may undermine the viability of conventional suction methods, unless it is possible to discharge directly to a dedicated lagoon or drying area. Lagoons need not be permanent.

To overcome the difficulties of increased water content whilst retaining the benefits of a closed pumping system, various attempts have been made to develop systems which employ fixed displacement pumps similar to those used for the pumping and placing of concrete. One such system is the EV dredger, where EV stands for 'environmental'. The system is of British design and manufacture and a prototype exists. The preliminary results of trials are encouraging, with pumped concentrations of 95% to 100% *in situ* material being recorded by independent assessors. Such performance is reliant on the *in situ* properties of the bed material and requires that a significant depth of debris free soft material is to be removed by dredging. Further development is in progress to widen the range of application.

Dilution of the soil by dredgers which excavate by bucket (see Sections 5.1 to 5.7), is far less than for conventional suction dredgers. Some dilution will nevertheless occur, the degree being dependent on the percentage filling of the bucket, whether the bucket has drainage holes and the depth below water of the dredged bed. Dragline and winch engine methods can generally be expected to dilute a little more than bucket, backhoe and grab dredgers. Subject to suitable precautions, such as bucket drainage, the percentage concentration of *in situ* soil can be expected to be in the range 75% to 95%, with 90% concentration (10% added water) being a reasonable expectation under normal conditions.

All of the preceding figures relate to added water or dilution of the soil. In all cases the soil *in situ* will also have a high natural moisture content, particularly unconsolidated sediments which are typical of maintenance dredging. The *in situ* density of highly organic material deposited in still water conditions in lakes, broads and some canals may be only 1.0 t/m^3, or even less. Non-organic deposits may have densities of 1.2 t/m^3, which approximates to only 12% dry solids. Mud deposits will rarely have a density greater than 1.5 t/m^3, or 30% dry solids, unless the mud has a high sand content.

Whether the soil water content is entirely natural, or increased by dilution, a reduction in water content, or thickening of the mixture, will simplify handling, reduce transport costs and improve the opportunities for beneficial use or disposal. Water content, or thickening, may be achieved by appropriate treatment (see Chapter 6).

5.11 RELATIVE COSTS

Economics are important in determining the right dredging method for a particular site. Whilst each method has its own physical and environmental advantages and disadvantages, economic attributes are less easily quantified. However, some basic influences can be identified, commencing with the initial mobilisation of dredging plant.

When mobilising floating plant via navigable waterways the cost is dependent on the length of travel and number of locks to be negotiated. If the route is short and free of locks, then costs may be competitive with land-based plant, but otherwise the slow rate of travel will result in higher costs. When access by water is not possible and transport is overland, mobilisation costs may increase dramatically, particularly if the dredger must be dismantled or launched by large cranes. For work requiring the dredging of only small quantities of material, the cost of mobilising dredging plant may be the most important economic factor influencing the choice of method.

The unit cost of dredging, that is the cost for each cubic metre removed, is highly sensitive to a wide range of site-specific operating conditions. It is not possible to provide meaningful relative unit costs for all methods of dredging.

The unit cost of dredging may be found by dividing the total hourly cost of the machine by the average hourly production. Total hourly cost may be the hire rate, or for an owned machine will comprise all of the cost elements, including fixed financial charges.

For machines which dredge mechanically by bucket, the average rate of production is dependent on the capacity of the excavating bucket in use, the average percentage filling of the bucket and the average cycle time. By combining these factors the average rate of production per effective hour can be estimated. Alternatively, production estimates for a dredger may be adjusted based on experience of the known time to fill a particular size of hopper barge under particular conditions.

Further adjustment will be necessary because a machine will not usually work effectively for a complete working day. Time will be lost in moving the machine, and in maintenance, re-fuelling, etc. It is usual to take effective hours as a percentage of the hours theoretically available in each day. The percentage of hours which are effective is highly site dependent.

Floating machines are usually more expensive than tracked or wheeled machines which work from the land. For example, though the basic machinery of a grab or backhoe may be identical whether on land or afloat, there will be some difference in the cost of working. Various factors contribute to the higher costs associated with floating plant, these include:

- higher capital costs arising from the pontoon and mooring system
- higher maintenance costs
- less easy access for crew and supplies
- higher insurance rates
- greater production loss due to extended machine movement time.

Assuming that in each case the method used is well matched to the dredging task, for machines of similar size or power, the relative production potential, ranging from highest to lowest, is as follows:

Potentially most productive	Water injection
	Bucket
	Cutter-suction
	Hydraulic backhoe
	Combined hydraulic-rope backhoe (VC)
	Dragline
	Winch engine
	Ploughing
Least productive	Grab

In suitable conditions ploughing, agitation and water injection methods may move material very cheaply, but none of these dredge in the conventional sense, in that they do not remove material from the system. The cost of using winch engines is heavily dependent on site characteristics. If the water body is of a regular shape with good access to one bank, which is not high relative to the water level, nor steeply sloping, and if securing a return block at regular intervals on the far bank is simple, then costs may be modest.

The basic cost of dredging, that is of simply removing material from beneath water, may only be a small part of the overall costs. Subject to site conditions, substantial additional costs may arise for treatment and disposal. As a consequence, the overall unit cost of dredging a polluted urban waterway may be many times greater than that of dredging a drainage channel through agricultural land, even though the plant used for dredging may be identical in each case.

Similarly, transport methods may have an important economic influence. Choice of a transport method may be dictated by the method of dredging, environmental considerations, or the place of disposal, and costs will be influenced by the isolation of the work site, the haul distance to the disposal site, physical restrictions to the route, and the sediment type. Transport should therefore also be considered early in the project assessment.

<u>Final decisions should only be based on production and cost estimates made by suitably experienced staff, consultants, or contractors.</u>

5.12 RELATIVE ENVIRONMENTAL IMPACT

Different dredging methods will impact on the environment in different ways. These have been described earlier in this chapter and each method has been given a relative environmental impact rating. For convenience, these ratings are summarised in Table 5.2. All of the ratings are subjective and assume normal use. A method with a high rating might be appropriate if operated with special care, which will usually mean with reduced production. Conversely, a low rated method might be harmful if used inappropriately.

A relatively high rating should not preclude the use of a particular method on an environmentally sensitive site. However, the implications of its use should be carefully considered. It may be that a particular method should only be used if subject to special constraints. For example, the use of lorries to transport dredged material over public roads might only be permitted, subject to the use of specially constructed vehicles, of a maximum permitted weight, within specified hours and with wheel cleaning facilities at the point of access to a road.

Table 5.2 *Relative environmental impact of different dredging and transport methods*

Plant type	Rating
Land based	
Winch-engines	10
Dragline	5
Hydraulic backhoe	4
Grab	3
Floating	
Rakes	7
Ploughs	6
Water injection	6
Bucket dredger - standard	3
Bucket dredger - modified	2
Cutter-suction dredger	1
Transport systems	
Road	8
Overland	7
Hopper barge	1
Pipeline	1

6 Treatment techniques

All treatment methods will add to the basic cost of dredging. Sophisticated methods may do so by several, or even tens of orders of magnitude. However, in some instances simple treatment, such as de-watering, may result in an overall saving in cost by facilitating less expensive methods of transport and disposal.

Treatment may be necessary or desirable for various reasons, but the most common are to reduce bulk, or the level of contamination, or to achieve separation. Reducing bulk, or weight, can result in savings in the cost of transport and disposal. Reducing contamination may render material acceptable for a wider range of disposal options, or uses, including options which are exempt from licensing under the Waste Management Licensing Regulations (DoE, 1994). However, a reduction in bulk may result in an increase in the percentage contaminant concentration, perhaps to the point where the material falls into a category of waste which may only be disposed of by special means and at high cost.

Separation may allow part of the dredged material to be disposed of beneficially or cheaply. For example, if material comprising a mixture of clay, silt and sand is contaminated, it is likely that the contaminants will be bonded to the fine clay or silt-sized particles. If the sand and coarse silt particles are separated, these may be disposed of more readily, leaving only a proportion of material to be disposed of by more expensive methods.

Most methods of treatment are applied after dredging and prior to disposal, but a few, such as bio-remediation, may be applied to the sediments *in situ*, usually as an alternative to dredging.

A difficulty with most treatment processes is that production rates are low relative to those of dredging plant. As a general rule, increasingly sophisticated treatment will be accompanied by a reduction in production rates and rising costs. The initial set-up cost of sophisticated or multi-process treatment systems may be high. For example, a permanent plant to separate sand from clay/silt/sand mixtures is under construction in Holland at the time of writing. The plant has a design throughput of 80 tonnes per hour and is estimated to cost approximately £1.6 million. In Belgium, a recently constructed permanent multi-process system for the processing of highly contaminated soils has a throughput of approximately 30 tonnes per hour and cost £10 million. Temporary low production installations can be provided at lower cost, but it is clear that all but the simplest treatment systems will be uneconomic on small-scale dredging works.

6.1 ROUTINE TREATMENT TECHNIQUES

6.1.1 Separation

It may be necessary to separate material which exceeds a specified size, or which must be disposed of separately, such as debris. A variety of methods are available including, coarse screens, trommels, skimmers and magnetic separators.

The object may be to remove metals or debris from otherwise clean sediment, or to separate contaminated fine particles from relatively clean coarser material. The latter may be necessary because contaminants are usually bonded to fine soil particles in the clay and silt range, though exceptions may be found.

Screens may be static, vibrated by a powered vibrator, or washed by pumped water. In each case, dredged material is discharged onto a horizontal or sloping screen. Undersized material falls through, or is washed through, into a receiving container or hopper. Coarse material is retained, or diverted, to an alternative receiver. Figure 6.1 shows a static screen separator. Figure 6.2 illustrates a powered screen separator.

In its simplest form, a screen may comprise of a coarse steel mesh grid fixed, or sliding, on top of a hopper barge, to prevent debris and oversize material from entering the hopper. Simple coarse screening may be necessary if material is to be disposed of on the bank or by spreading on agricultural land. Unless the area to be dredged is highly polluted by debris, the use of simple fixed screens should not have a very significant effect on productions rates or costs.

Whereas simple screening may be effective in removing debris, it will not be fully effective in separating fine from coarser materials if the difference in particle size is small, or if the material comprises a wide range of particle sizes. If screening is inappropriate, other methods, such as washing (see Section 6.2) or hydrocyclones, may be necessary. Dutch experience shows that the separation of sand by hydrocyclone is uneconomic for sand contents below 40%.

Figure 6.1   *Static screen separator*

Figure 6.2 *Powered screen separator*

The sand and gravel industry, and also the metal and coal mining industries, have developed a wide range of separator and screen types, each suited to a particular function. Suppliers to these industries can provide standard screens of various types, together with a range of complimentary equipment, such as stacking conveyors, hoppers and separators. Much of the equipment is designed to be portable and for open site use, and is suitable for use on dredging works without major modification.

The influence of separation methods on overall cost is dependent on the sophistication of the process. The following list of options is arranged approximately in ascending order of cost:

Least expensive	Fixed static screen
	Washed screen
	Single-stage vibrating screen
	Multi-stage vibrating screen
	Up-washing, or boiling box (see Section 6.2.1)
	Multi-stage vibrating screen with magnetic separator
Most expensive	Hydrocyclones (see Section 6.1.2)

6.1.2 De-watering

It is usual for dredged sediment to have a very high water content. This may complicate transport, disposal, or eventual use, hence it is often necessary, or at least advantageous, to reduce the water content (see Figure 6.3). This results in reduced bulk and weight and, in most cases, in easier handling.

The operators of commercial licensed waste disposal sites will usually refuse to accept waste with a high water content. The definition of 'high water content' may vary from site to site, but generally if the material is 'sloppy', such that handling and spreading is made difficult, the material may be rejected, or subject to a substantial cost surcharge.

The ease or difficulty of de-watering is dependent on soil characteristics. Granular materials, including sands and gravel, will drain quickly whilst the release of water from clays, silts and mud will take much longer. The drainage of fine grained material may be improved by mixing with granular soil or reworking.

The simplest method of de-watering is by a combination of drainage and evaporation (see Figure 6.4). Dredged material which is spread thinly over land will usually drain quickly, particularly if the land is free draining. As the thickness of the deposit increases, paths of drainage and evaporation are lengthened and drying will take longer. When fine grained material is deposited with a thickness in excess of one metre they may retain a high moisture content and exhibit low strength for a long time, such that reworking, or beneficial use, is difficult.

If fine grained dredged material is to be disposed of within purpose built containment areas, especially when placed by pumping to lagoons (see Section 5.9.4), it is important to limit the depth of deposition at any time. De-watering will be most rapid if the disposal site is sub-divided and filled on a rotational basis. Each sub-area should initially be filled only to a depth such that adequate drying can occur before the next stage of filling commences. Interim reworking of the soil by ploughing or rotavating may be necessary. If feasible, it will be helpful if intermediate thin sand layers are pumped over the area prior to placing the next layer of fine grained material. By these methods, an eventual deposit thickness of several metres may be possible, whilst achieving an acceptable overall moisture content and strength.

More sophisticated de-watering methods are available. These include centrifuging, pressing, or the use of heat. All are expensive and generally suffer from low throughput relative to the production output of dredging plant. Hence their adoption may be limited to special situations. However, the use of hydrocyclones, particularly in association with hydraulic (e.g. cutter-suction) dredging methods, has been quite widely adopted in Belgium and Holland and on occasions in the UK (e.g. the British Waterways 'Aquarius' project in Birmingham). In contrast with other methods, relatively high throughputs are possible without excessive energy input. The method may be used for thickening or separation.

The relative costs of de-watering methods are influenced by production rates and energy consumption. The following list of options is arranged approximately in ascending order of cost:

Least expensive	Open field site drainage and evaporation
	Contained site drainage and evaporation
	Contained site drainage and evaporation with re-working
	Hydrocyclones
	Belt pressing
Most expensive	Heating

Figure 6.3   Wet material contained within a shallow bunded enclosure

Figure 6.4   Shallow bunded enclosure after drainage and evaporation

Thickening may also be achieved by the use of additives, such as lime. Dredged material and additives are mixed in a drum mixer. The thickened material is more easily handled, which may simplify transport and render the material more acceptable to the operators of licensed waste disposal sites, or into a condition suitable for beneficial use.

6.2 SPECIAL TREATMENT TECHNIQUES

6.2.1 Washing

Washing may be used to separate fine grains from coarser material. Washing may be undertaken in conjunction with screening (see Section 6.1.1) or as an independent process. In the latter case washing is achieved by use of a 'boiling box', or similar arrangement, wherein the dredged soil mixture is discharged into a container of up-flowing water; also known as the 'fluid bed' method. Coarse grains fall through the water column whereas fine grains are suspended and removed by the discharge water. Subject to the size of the plant, the approximate particle size at which the split is achieved can be adjusted by varying the rate of water flow.

A refinement of the washing process is known as 'frothing' (see Section 6.2.2).

The relative costs of different washing methods are very dependent on the method of dredging adopted and particular soil conditions. When a hydraulic method of dredging and soil transport is employed, the pump discharge pipeline may be fed directly to a boiling box. Fine grains may then be separated with only a small increase in costs. In contrast, if a mechanical method of dredging is employed, the use of a boiling box may increase costs substantially. For mechanical dredging methods, it may be cheaper to achieve separation by direct discharge to washed screens, although the minimum particle size at which separation may be achieved is likely to be larger with this method.

If the original material is contaminated, most of the contaminants may be removed with the fine material, but some may be transferred in solution to the wash water. The system must then be closed so that the contaminated water does not pollute ground water, watercourses, or the environment in general.

Subject to the type of contamination, the quality of the drainage water may be improved by trickle discharge via a treatment area comprising reed beds. Enriching chemicals, such as phosphates and nitrates, can be effectively removed or reduced by this method (see Section 6.2.2 and CIRIA, 1996b).

The discharge of water from any drainage or washing process requires a consent to discharge from the Environment Agency.

6.2.2 De-contamination

The treatment of contaminated material may be expensive compared to isolation, which only requires that the material to be disposed of is isolated from the general environment, as in landfill. The available methods for de-contamination may involve the addition of substantial volumes of chemicals, or substantial energy input in the form of heat. Such disadvantages must be weighed against the positive environmental benefits. For these reasons it will usually be the case that de-contamination treatment is reserved for only severely contaminated materials.

For mildly, or moderately contaminated materials, isolation or biological treatment are most likely to be cost effective and when viewed overall may have the least impact on the environment.

The relative costs of different methods of treatment for contamination are highly dependent on the properties of the material to be dredged and the particular contaminants. It is, therefore, difficult to provide general guidance concerning relative costs. However, it will **usually** be the case that for the following list of options costs will increase in an approximately ascending order:

Least expensive	Biological
	Chemical
	Immobilisation
Most expensive	Thermal.

Biological methods

These methods divide into two quite different approaches, those which rely on uptake by growing plants and those reliant on the action of micro-organisms (Detzner, 1993).

Plant uptake methods subdivide into plants which collect contaminants via their root system and algae which adsorb contaminants through their cell walls. A range of plants, including the reed/rush family, are known to collect contaminants, though their ability to do so varies with plant type and is also dependent on the characteristics of the sediment and contaminants.

These methods may be employed in various ways. Water draining from contaminated dredged material may be routed through an enclosed existing or purpose-built area cultivated with appropriate plants. The total capacity of the dedicated area must be matched to the rate of water throughput.

Alternatively, appropriate plants may be grown directly on deposits of dredged material with the object of removing water by transpiration. For this process to be fully effective the total material thickness must not exceed the depth of root penetration. An example of an application for which such methods may be suitable, is where mildly contaminated dredged material is used to construct wetlands. Following the deposition of the dredged material to create the required wetland conditions, appropriate planting, or sowing, may serve the dual purposes of cleaning and stabilisation (see CIRIA, 1996b).

Chemical methods

Frothing is a refinement of the washing process. Frothing chemicals are added to the mixture which is subject to air entrainment. Contaminants and very fine sediment particles are trapped in the froth which is skimmed off for further treatment or disposal.

Contaminated material may also be treated chemically, and again methods may be divided into two broad categories depending on the objective, which may be either to neutralise or to remove the contaminants.

Organic contaminants may be neutralised by mixing with oxidising agents, such as peroxide. However, the necessary quantity of added agent may be large and hence expensive.

Polychlorinated biphenyls (PCBs) may be neutralised by mixing with sodium or potassium which react with the chlorine atoms.

Chemical treatment methods are currently very expensive and hence restricted to use on exceptionally harmful materials. New methods are being developed which hold the promise of lower costs. One such method utilises the ability of microbes to produce sulphuric acid when consuming organic compounds to remove heavy metals, except lead (Schotel and Rienks, 1993).

Immobilisation

Immobilisation methods depend on either solidification, such as conversion to bricks, or fixation, whereby chemicals are added to 'fix' the contaminants to the soil structure. Various chemical methods exist, for example, by the use of hydroxyl forming agents or a silica solution. The long-term stability of fixation methods is uncertain and hence may not be acceptable in all situations. Solidification may provide an effective immobilisation of contaminants, but is expensive and material dependent.

Immobilisation methods are appropriate only as a last resort for highly contaminated materials.

Thermal methods

Thermal methods require substantial energy input and hence have the dual disadvantages of being expensive and environmentally flawed. In common with immobilisation, they represent a last resort only applicable to seriously contaminated materials when a more satisfactory method is not available. Thermal methods include:

- desorption
- incineration
- thermal reduction
- vitrification.

Discussion of these methods is beyond the scope of this guide. Readers are referred to PIANC (1996) and specialist papers.

7 Contractual matters

The work of inland dredging may be performed by the owner of a waterway, or the responsible authority, using internal resources of plant and labour. This is known as the 'Direct Labour' method. However, increasingly work is performed by a Contractor, usually a specialist, in which case a form of contract is required which defines the responsibilities and obligations of the parties to the contract. The contract must also specify in detail the work to be performed and the methods of measurement, valuation and payment. The party ordering the work is usually described as the 'Employer' and the party executing the work as the 'Contractor'. The work may be supervised by a member of the Employer's staff, or by a third party, usually a consultant retained by the Employer. In engineering contracts, the supervisor is termed the 'Engineer'.

7.1 FORM OF CONTRACT

If works are to be performed by a Contractor, rather than by direct labour, a form of contract between the parties will be necessary.

There are many forms of contract, standard and non-standard, and many methods of measurement which might be employed. In this document, those considered to be good practice are described.

The contract provides the legal framework within which the work is to be performed. It is, therefore, essential that the form of contract be well understood by all parties to the contract and well tested in law. These basic requirements point unequivocally to the use of a nationally recognised form of contract, such as ICE 5th (1979) or 6th (1991), or for work of modest scale, ICE Minor Works (1988).

When the use of such a form of contract is adopted, modification of the terms should be resisted so far as is possible. Modifications may cause confusion, or may be unintentionally misleading. They may alter the effect of unaltered clauses and, most importantly, unlike the original form, the modified form may be untested in law.

Input to the drafting of national forms of contract is made by all sides of the industry. The final form is therefore reasonably well balanced. In contrast, forms of contract which are drafted only by the Employer almost invariably will be weighted in favour of the client, perhaps to an unreasonable degree.

A reasonable Employer will recognise that it is the role of the Contractor to provide a service whereby the Contractor performs a specified task to a quality, within a defined period and at a cost, which is reasonable under the circumstances that prevail during the execution of the work.

If the circumstances change to a degree which could not reasonably be foreseen by the Contractor, and that change renders the work more difficult, the Contractor might reasonably expect to receive additional payment. This may cause the Employer some difficulty, particularly if constrained by tight budgetary constraints. This is a problem for the Employer, but is no basis for resisting reasonable payment to the Contractor.

7.2 SPECIFICATION

A specification provides the Employer with the means to describe to the Contractor the work which is to be performed, the quality which is to be achieved and any constraints on method, practice, or working time, which are to be imposed.

When drafting a specification, it is important that the draughtsman recognise that his or her understanding of the requirements may be highly developed. Understanding of the site characteristics and environment is also likely to be good. In contrast, the Contractor cannot know what is in the employer's mind and before tendering may have never seen the site. This can easily result in an unconscious assumption by the draughtsman that the Contractor knows what he or she knows, which can lead to a specification which is incomplete, ambiguous, or misleading and this should be guarded against.

It is usual practice to divide the specification into a general specification, which describes general matters such as access to the site and records to be kept, and a particular specification, which details the work to be performed and any constraints which may significantly affect the Contractor's operation. Division into general and particular matters is not essential; the two may be combined when the proposed work is very simple.

Each specification should be drafted specifically for the intended work. However, for inland dredging, certain features are likely to be common to most works. A specification, therefore, should include descriptions, as appropriate, of the matters listed in Boxes 7.1 and 7.2.

Box 7.1 *General specification*

- Location of site.
- Extent of work.
- Sequence of work.
- List of drawings.
- Units to be used, i.e. metric.
- Datums to be used.
- Access.
- Conditions concerning navigation, locks, sluices, etc.
- Security, safety of public, exclusion of farm stock, etc.
- Signs to be used.
- Temporary site offices, work and storage areas, etc.
- Provision of facilities for Engineer, if any.
- Requirements for keeping records of production, plant, labour, progress, and photographic records
- Requirements for obtaining and testing samples of water and sediment.
- Safety and welfare requirements.
- Requirements for the protection of services; power, water, telephone, bridges, culverts, etc.
- Requirements for the protection of trees, hedges and vegetation.

Box 7.2 *Particular specification*

> - Work to be performed (a detailed description should also be provided as an introduction to the Contract documents).
> - Material to be dredged.
> - Method of dredging (if necessary; where possible it may be better to leave the choice to the Contractor).
> - Method and place of disposal of dredged material (choice may also be left to the Contractor).
> - Control of water levels.
> - Records of the disposal activity, load number, weight, time and place.
> - Method of treatment, if any.
> - Sequence of dredging, if necessary.
> - Dimensions.
> - Slopes.
> - Tolerances.
> - Effect of natural siltation, if significant.
> - Restrictions on the spillage of sediment and water from hoppers, trucks, or pipelines.
> - Limits on induced suspension or dispersal of suspended solids.
> - Water quality in vicinity of dredging and drainage from lagoons.
> - Environmental objectives and working practices.
> - Precautions against pollution and action in the event of spillage of oil, chemicals, etc.
> - Method of measurement of water level where variable and if critical.
> - Setting out, key points, bench marks to be used, etc.
> - Temporary movement of navigation buoys or marks.
> - Restrictions on over-dredging.
> - Dredging in the vicinity of structures.
> - Priority of navigation.
> - Speed limits for navigation or vehicles.
> - Due care in respect of bank or fishermen.
> - Permissible noise levels.
> - Working hours.
> - Method of proving satisfactory completion.
> - Reinstatement.

7.3 METHOD OF MEASUREMENT

As in most engineering works, it is usual that payment for dredging is based upon a particular form of measurement of the works. Measurement is made initially to determine the expected extent and nature of the work and, finally, to determine the quantity and quality of the work done. Usually the particular method of measurement adopted will be influenced by the nature and scale of the work.

7.3.1 Measurement for payment under contract

The most common methods of measurement of inland dredging work are described here. Whichever method is adopted, the method and details of measurement should be clearly described in the Preamble to the Bill of Quantities. The Bill of Quantities is an itemised schedule of each type of work to be done, with the approximate quantity of work given against each item. Specialist advice should be obtained unless the Employer's staff have substantial and sound experience of the methods specified. Selection of the most appropriate method is strongly influenced by the scale and nature of the particular work to be done.

For the maintenance of a land drainage channel of more or less uniform profile requiring a fairly uniform amount of work per unit length, simple linear measure may be most appropriate.

In contrast, the restoration or creation of a navigation channel in a major river, or through a broad, may involve a substantial volume of dredging with wide variations in length. In such cases measurement of volume is likely to be most equitable.

If linear measurement is adopted, it is necessary to provide those who are to submit prices for the work with sufficient detail to allow a reasonable assessment of the amount of work to be done. This is most easily achieved by providing a set of typical cross-sections, showing the existing and required profiles and tolerances, if any. If the work is reasonably consistent in length, only a few typical sections may be necessary. As always, the persons pricing the work should be required to visit and examine the conditions at the work site. The work will be satisfactorily completed when it is shown by measurement that the required profiles have been achieved, or are within tolerance (see below), throughout the length of the work.

If volumetric measurement is adopted, it is necessary to provide more detailed information concerning the existing conditions of the site. This may take the form of cross-sections or plan charts showing levels and contours. The necessary spacing of cross-sections is dependent on site conditions, particularly the variation in section with length. Canals will usually be reasonably uniform, but rivers may exhibit wide variations in section, particularly between straight reaches and bends. It is not appropriate to specify a particular spacing here, but as a general guide, spacing of 15 m might be appropriate for channels of widely varying section, whereas 100 m spacing might be satisfactory for channels of fairly uniform section.

When channel section is generally uniform, but with occasional substantial variation, such as at 'winding holes' on canals, it may be better to provide for different and separate measurement of such areas.

For large or irregular areas of water, a plan area, rather than end area method of computing volumes (see Section 7.3.2), may be more appropriate. The plan area method requires that points of equal level be joined by contour lines. The plan area may be measured by planimeter, or arithmetically, and multiplied by the appropriate depth, usually the vertical interval between adjacent contours, to give a volume. Subject to the particular circumstances, it may be more accurate to multiply the vertical interval by the mean of the areas enclosed by adjacent contours.

Tolerances

When drafting a contract based upon sectional or volumetric measurement it is necessary to decide whether or not over-depth tolerance should be specified and whether material removed from within the tolerance should be measured for payment.

Over-depth tolerances may be specified to protect structures, or in recognition of the fact that it is impossible to achieve a particular specified bed level without a margin of over-dredging.

Tolerances, whether vertical or horizontal, which are necessary to avoid undermining structures or banks must be specified.

Tolerances which recognise the necessity of some over-dredging are optional, but it should be recognised that some over-dredging is necessary to achieve a specified level and must be paid for, whether by direct measurement, or whether included within composite rates tendered by the Contractor. The necessary working tolerance is dependent on the type of dredger to be employed; an auger cutter-suction dredger can work within a smaller tolerance than a grab dredger. Factors which will influence the working tolerance are listed as follows:

- type of dredger
- size of dredger
- operator skill
- bed material
- water depth
- site conditions (variable water level, vegetation cover, wave height in open waters).

It is important to recognise that the specification of a very small tolerance may substantially reduce the rate of production of the dredger and hence result in higher unit costs. When specifying a vertical tolerance for navigable waterways, it is usual to require that all material is removed from above a specified level. The plus tolerance is then zero. Working tolerances which are easily achieved by inland dredging plant are shown in Table 7.1.

Table 7.1 *Working tolerances for inland dredgers (all plus 0, minus tolerance)*

Type	Small in still water mm	Medium in open water mm
auger cutter suction	70	100
crown cutter suction	100	150
bucket dredger	150	200
backhoe	150	250
grab	200	350
dragline	250	500
winch engine	500	750

There are advantages and disadvantages attached to measuring for payment quantities dredged from within an over-depth tolerance. It is the authors' experience that for the Employer, the disadvantages are greater than the advantages and hence it is not recommended that over-dredging be measured for payment. Measuring and paying for over-dredging suffers a range of disadvantages, as follows:

- quantity to be paid cannot be known at time of contract award
- use of less appropriate plant may be encouraged
- use of inexperienced operators may be encouraged
- reduced incentive for good supervision and control by Contractor
- larger volume than is necessary may be removed
- larger volume for disposal may result
- bank instability may result
- greater environmental impact may result.

The only advantage for the Employer of measuring and paying for over-dredging is that the tendered rate for dredging is a true rate per cubic metre rather than a composite rate, which includes the Contractor's estimate of necessary unpaid over-dredging. In some situations this might adversely affect the valuation of variations. For example, if the work initially priced by the Contractor is to remove an average depth of 500 mm of bed material, he or she may allow for the removal of a further 100 mm depth which is not measured for payment. This is done by increasing the tendered unit rate by 20% relative to the real estimated rate. If the Employer subsequently issues a variation order instructing the Contractor to dredge deeper and remove an average of 1,000 mm, the increase in the rate needed by the Contractor will fall to 10%, but the original tendered rate will be used in valuation, unless a new rate is negotiated.

Contracts without measurement, other than to check for compliance

The methods of measurement described all assume one-off contracts. For channels which require regular maintenance, consideration should be given to term contracts, by which the Contractor undertakes to maintain a specified depth, or profile, for fixed annual or stage payments. Under such arrangements, no re-measurement, other than checking compliance, is necessary.

Fixed payment term contracts are only possible where an adequate knowledge of the average amount of work necessary exists. Such knowledge requires a minimum of 5 years of good quality survey and dredging information.

7.3.2 Survey methods for measurement

The discussion in Section 7.3.1 assumes that measurement is by conventional cross-sections surveyed by land-based levelling, or water borne echo sounding.

Land survey methods

Land-based methods usually employ a level, set up on one bank of a channel, and a levelling staff (see Figure 7.1). Measured levels are reduced to Ordnance or a local datum. If used in a channel with a soft bed it may be necessary to attach a circular plate to the base of the staff to avoid penetration below true bed level. The appropriate plate diameter is dependent on the consistency of the bed material, but a minimum of 100 mm is recommended.

As an alternative to direct levelling from a waterway bank, measurement may be made by dipping with a graduated staff to measure the depth from water level to bed level at intervals across a section. It is then only necessary to determine the water level by levelling relative to Ordnance Survey bench marks, or other recognised datum. If the water level is variable, it is necessary to measure carefully water level at regular intervals throughout the period of survey and simultaneously to record the time of all measurements made relative to the water level.

Figure 7.1 *Simple, but laborious and unsafe way survey method (surveyor should be connected by a line to the bank)*

If measurement is by land levelling or dipping, the points listed in Box 7.3 are important:

Box 7.3 *Check list for levelling or dipping surveys*

- All levels should be related to a known datum, preferably Ordnance Datum.
- Reliance should not be placed on a single bench mark, unless previously checked for accuracy.
- Horizontal measurement at sections should be relative to an established baseline or feature.
- If the bed material is soft it should be decided what is to be measured, i.e. the level of the top of sediment, or the firm base layer, or both (usually preferable). Care should be taken to adopt methods which accurately measure the chosen feature or features.
- Flexible measuring equipment, such as sounding lines, should not be used if current and water depth are such that significant deviation from the vertical measurement might occur.
- If dipping relative to water level, measure water level relative to Datum. If water level is variable, measure level through period of survey.

Waterborne survey methods - hydrographic survey

Echo sounding is conducted from a boat. The echo sounder transducer is submerged to a known depth and the instrument generates a continuous series of sound impulses which are transmitted through the water column and reflected from the channel bed. Circuitry within the instrument measures the return travel time of the acoustic pulse and converts this to a distance. The depth of water below the level of the transducer is recorded on a paper chart, or magnetically on a disc. If it is required to measure both the surface of soft sediment (bottom of the water column) and the level of firm underlying strata, a suitable dual frequency echo sounder or pinger should be used.

If using echo sounding methods, the following points are important:

- use only a survey quality echo sounder, not a model designed for fishermen
- if digitising output, also record an analogue graphical record for checking
- if digitising output, check for selectivity (e.g. are depths averaged, or least depths selected?)
- carry out 'bar-check' to adjust speed of sound in water (varies slightly with temperature and salinity)
- do NOT significantly alter 'gain control' subsequent to bar-check
- do NOT adjust 'draught control' subsequent to bar check
- do NOT alter draught or trim of survey boat through excess use of power or movement of load
- on conclusion, repeat bar check
- measure water level relative to Datum; if water level is variable, measure level throughout period of survey
- for long lines, as in lake or reservoir surveys, subject to the method of positioning used, the speed of the boat should be approximately constant over the full length of the line.

Volumes are computed by assuming that each cross-section is typical of a length equal to the distance from adjacent cross-sections. This is known as the 'end area method'. Measurement of cross-sectional area at each section may be by planimeter, or arithmetically by Simpson's Rule.

For maximum accuracy, or for large areas, the best results will be achieved by digitising the bathymetry and using computer ground modelling methods, sometimes referred to as digital terrain models, digital ground models or digital elevation models.

New and rapidly developing technologies, such as multi-beam echo sounding, sonar and ground radar, provide a means by which the continuous rapid measurement of bed levels is possible in and adjacent to areas where water depth is sufficient for navigation. Such systems offer a range of possible advantages including, rapid area coverage, 100% bottom cover, measurement of side slopes to water level, measurement of any undermining of structures and, with some systems, the separate measurement of the levels of superficial soft sediments and underlying clay lining or virgin ground (see Section 10.1).

In addition, continuous acoustic or radar methods, when allied with suitable ground modelling software, provide a choice of presentation in plan, cross-section, or three-dimensional projection. Ground modelling software allows rapid computation of volumes and hence a range of design profiles and alignments can be quickly assessed. Such systems, when fully developed, will be well suited to measurement on larger projects, or where the routine monitoring and analysis of long lengths of channel is required.

Accurate positioning within open waters, such as lakes, reservoirs, large rivers and canals, is difficult and will be best achieved by use of Differential Global Positioning Systems (DGPS). To achieve a satisfactory level of accuracy it will usually be necessary to establish a temporary local differential station. In general, the accuracy of DGPS is improving rapidly, but systems currently in common use may not be sufficiently accurate for the detailed survey of narrow waterways, or waterways in areas with poor DGPS reception. Highly accurate systems are available, but at present may be too expensive for small firms or infrequent users. As in computing, the ratio of power to price of DGPS systems is improving rapidly and it can be assumed that the use of DGPS will soon be routine for both hydrographic and land based survey work.

All water-borne systems of survey may become less economic if navigation of the length to be surveyed is frequently restricted by locks, or areas too shallow for navigation.

If a large-scale hydrographic survey is planned, as for lakes and reservoirs, reference should be made to BS 6349, Part 5 and *Guidelines for the preparation of hydrographic survey specifications*, published jointly by ICE and RICS (1983).

7.4 OBLIGATIONS, LEGAL AND CONTRACTUAL

The obligations on Employer and Contractor respectively are defined by the contract. Provided that well-established and widely recognised forms of contract are employed, the matter needs no detailed discussion here. However, some issues are occasionally the subject of misunderstanding.

Provision of information

Those who are to undertake the work will usually gain knowledge of the site conditions through information supplied by the Employer, or his or her consultants. The work may be undertaken by an in-house Direct Labour organisation or by a Contractor. It is assumed here that work is by Contractor, but the underlying requirements may be the same for both.

Only with the benefit of adequate and reliable information can a Contractor assess and estimate the amount of work to be done, the conditions under which it must be carried out, the rate of production which might be achieved and the cost of achievement. Usually, if work is awarded by competitive tender, contractors will have only a short time, typically a few weeks, for the preparation and submission of tenders.

It should be clear that for all but the simplest of tasks, time will be too short to permit tenderers to investigate the site, collect and analyse information and reach a level of knowledge and understanding which is comparable with that of the Employer. This might be overcome by allowing a much longer time for tendering, but this will result in all tenderers having to devote resources to the detailed investigation of a project which ultimately only one will be asked to carry out. This is not satisfactory. It will be better if the Employer, or his or her consultant, gather and provide adequate and proper information to all tenderers. It follows that the Employer is then responsible for the accuracy of the information provided. However, it is not uncommon for the Employer to disclaim responsibility for the information provided, in the apparent belief that by so doing he or she will be protected against the risk of increased costs in the event that the information provided is found to be misleading. <u>Numerous cases in law have shown this to be a misconception.</u>

Legality aside, the ethics of disclaiming responsibility for information provided are questionable. It follows that the responsible Employer will take all reasonable measures to provide appropriate and accurate information when possible. The information most important in the assessment of inland dredging works will usually include the following:

- quantity of material to be dredged and disposed
- method and place of disposal
- quality to be achieved by treatment, if any
- dimensions, including water depth
- characteristics of the material to be dredged
- nature and density of foreign matter, including debris
- access
- constraints on operations, i.e. working hours, water quality, navigation, etc.
- season during which work is to be performed.

7.5 CONTRACTOR SELECTION, TENDER APPRAISAL, CONTRACT AWARD

Selection

For public authorities it will usually be necessary to advertise in the local and national press the intention to invite tenders for all substantial works. It may also be necessary to advertise in the *European Journal* (Supplement to the official journal of the European Communities, EUR-OP, 2 Rue Mercier, L2985, Luxemburg). The threshold values at which advertising becomes necessary vary and enquiry should be made within the legal administration of the authority in question to determine the current level appropriate to the work proposed.

It is important to invite tenders only from companies who are competent to carry out the work. This implies that invitations be issued only to a select list of companies whom investigation has shown to have plant, experience, organisation and financial strength which are appropriate to the type and scale of work to be performed. There are many dangers attached to an open tender approach, whereby all and sundry may submit a tender for the work. For public authorities, the greatest danger is that the lowest price may be submitted by a company which is not competent to undertake the work. The officers of the authority may then be faced with the sometimes difficult task of convincing a lay committee that the lowest price should not be accepted. It is far better that only those companies to whom the work can be awarded with reasonable confidence be invited to submit tenders. Open tenders also waste resources within the organisations of both Employer and Contractors.

The danger of the select list approach is that innovative and emerging companies may be denied the opportunity to develop. This risk can be avoided by initially encouraging such companies to tender for works of modest scale and complexity and to progressively raise the level of work for which they are invited to tender in line with their proven capability.

Adequate time should be devoted to the selection of tenderers. It should be a relatively simple matter to determine financial strength and ownership, and access to appropriate plant.

Competence and experience are less easily established. The usual method is by taking up references. However, written references are not always reliable. A more thorough understanding is likely to be gained by the selection of one or more previous employers of a Contractor for whom broadly similar works have been performed and by visiting these employers to discuss their experience face to face.

Alternatively, the informal views of a suitable specialist consultant may be obtained, though individual consulting firms may not have adequate knowledge of all companies who may apply for pre-selection.

Tender appraisal

The appraisal of tenders received from a select list of contractors is simplified because, if the initial selection was good, the award may simply be made to that Contractor who submits the lowest tender. However, it is first necessary to determine that the Contractor who has tendered the lowest sum is most likely to complete the work at the lowest cost, within the specified time and to the required quality. This does not necessarily follow.

For all but the simplest of works, the sensitivity of final cost to variations in individual bill items should be checked. For most works there will be areas of uncertainty. These may be in measured quantities, or payment for delay. Best estimates may have been used in the Bill of Quantities, but what if these are wrong?

The range of possible variation should be considered and the minimum and maximum values applied to the tendered rates to determine the effect on the total price. This is most easily done by constructing a simple computer spreadsheet which mirrors the Bill of Quantities. The rates of each tenderer are then entered and the effect of variations in quantity or time tested.

It is not unusual for an experienced Contractor to spot an item for which the billed quantity may be too low and to enter against that item an inflated rate. If the billed quantity is small, the effect on the total tender price may also be small and hence the effect on the competitive placing of the tender may be only marginal. If the Contractor's judgement proves correct and the true quantity is much greater, then the total price of the work may increase substantially and legitimately and beyond the control of the Employer.

The method of testing sensitivity by use of a spreadsheet also serves the dual purpose of checking the arithmetical accuracy of all tenders.

The tenderer's ability to comply with the contract programme is less easily checked. It requires study by a person with appropriate experience of similar work and methods, or by comparison with other completed works of a similar nature and which employed methods similar to those proposed by each tenderer. As individual tenderers may propose different methods, the Employer may lack the experience and the information necessary to adequately assess whether or not a contractor's assumed rates of production are realistic. In such cases it is recommended that specialist advice be obtained.

Quality will be influenced by the contractor's attitude and organisation and by the plant and methods to be employed.

If all tenderers have been pre-qualified, then it may be assumed that their performance and organisation are satisfactory, but it will be prudent prior to award of contract to determine that the experience of the particular key personnel who are to be employed on the works is appropriate and adequate.

The standard of quality which can be expected of particular plant and methods may require experienced assessment. Some differences should be fairly obvious. For example, if it is important that dredging be to a particular level, with only small permitted tolerance, then a floating grab dredger cannot be expected to achieve the same quality of result as a cutter suction or bucket dredger. Similarly, a land-based grab, or dragline, is likely to produce results inferior to a hydraulic backhoe (see Chapter 5). The implications of different plant and methods are too complex to detail here. What is important is the recognition that different plant and methods may, and probably will, produce different results and the cheapest may not be the most satisfactory.

Finally, if not already considered under quality, it is necessary to assess the relative environmental impact of the various plant and methods which may be proposed. Environmental aspects should be an integral part of any contract and contract measurement method, and should be formalised.

Operations may impact adversely on water quality, bank-side flora, agriculture, other users of the waterway, the community at large, the atmosphere, the place of disposal, etc.

Impact due to noise may be reduced simply by restricting working hours, but also by rapid completion. A Contractor whose proposed methods are cheap, but relatively noisy, and which require twice as long for completion as other methods, is unlikely to be welcome within a residential community. Similarly, a method which results in a sharp deterioration in water quality, but for a very short time, may be preferable to a method with lower maximum impact extending over a long period, particularly if by taking longer the work overlaps with spawning or breeding seasons.

The assessment of relative environmental impact may require very experienced judgement, perhaps with input from different disciplines. Simple guidance has been provided for each basic type of plant and method in Chapters 5 and 9, but in complex situations this may not be sufficient for a proper assessment.

Contract award

Following completion of the various analysis and assessments described in the previous sub-sections, the award should be made to that Contractor who emerges as the most likely to complete the work at the lowest cost compatible with the required quality and programme and with the least adverse long-term impact on the environment.

7.6 CONTRACT SUPERVISION

Dredging is a specialised operation, complete knowledge of which is dependent on a combination of theory and relevant experience. Because much of the work to be performed is hidden, being submerged and buried, unforeseen events are common. Such circumstances require a pragmatic and flexible approach to contract supervision.

Supervision should be by a person with a reasonable understanding of the plant and methods which are to be employed in the execution and measurement of the works. Inexperienced individuals should not be entrusted with the supervision of contract works, unless their primary function is simply that of providing eyes and ears, with important judgements being entrusted to an appropriate superior.

The appropriate level of supervision will be influenced by the scale and complexity of the works, with larger projects justifying more experienced supervision. However, it may be more difficult to complete a small project satisfactorily. This is because, with little time and money available, efficient operation must be rapidly achieved and any problems arising must be dealt with promptly. These problems are greatest for the Contractor, but the representatives of Employer and Contractor should share common objectives; the efficient and satisfactory completion of the work.

A common objective approach to contracts is very important. Of course the representatives of Employer and Contractor may have different priorities. The Employer wishes to maximise quality and to have the work completed within budget. A good Contractor will also wish to produce work of good quality and make a profit. If for the Contractor to make a profit, payment for the work exceeds the Employer's budget, this will be a potential source of conflict, but the risk of serious conflict will be minimised if both sides co-operate at all times in seeking the optimum solution to any problems which may arise. Both sides should beware of getting into entrenched and opposing positions.

Contractual disputes are commonly rooted in personal conflict. A sensible, flexible and pragmatic attitude by the key supervisory staff of the major parties to a contract, will usually result in satisfactory completion of the work at a reasonable price, even if the wording of the contract is imperfect. In contrast, a well-drafted contract is no guarantee of satisfactory completion when the personalities of key supervisory staff are incompatible, or if one or both lack adequate and relevant experience and understanding of the work.

It is common for marine dredging operations to be continuous. That is, operational for 24 hours a day, 7 days a week. This is less common for work inland, but may occasionally be necessary on substantial projects which employ floating plant. Clearly if work is continuous, both Contractor and Employer may need to extend supervision beyond normal working hours. This is most important for the Contractor, who should respond rapidly to unforeseen events, such as the breakdown of key plant. It is unlikely that it will be necessary for the Employer to provide 24-hour attendance, except in the most complex or sensitive situations, but 7-day attendance might be necessary. Alternatively, it may be sufficient that a representative of the Employer be on call.

As well as issues relating to work practices, measurement, problems, etc., part of the contract supervision role is to ensure that the environmental objectives and constraints included in the contract specification are adhered to. These should be given equal weight as other engineering and contractual matters.

The quality of supervision will be usually more important than the quantity.

7.7 RESPONSIBILITIES

In all works of civil engineering which are performed under contract some responsibilities rest with the individual parties and others are shared. The level of duty ranges from a strict legal requirement to that which is merely desirable or prudent. No distinction is made here between different levels of responsibility. The responsibilities which are particularly relevant to the work of inland dredging are outlined in Boxes 7.4, 7.5 and 7.6.

Box 7.4 *Duties of the Employer*

- To describe and, if possible, identify any potential hazards, including: live power cables, pipes carrying oil or gas, ordnance, weak or vulnerable structures.
- To accurately describe the physical conditions of the site and any constraints on working.
- To inform the public of the object, nature and effect of the works.
- To prepare a safety plan and maintain a safety file and appoint a Safety Supervisor, all in accordance with CDM (see Chapter 8).
- To obtain all necessary consents and/or permissions, including:
    - to obtain planning consent
    - to enter, occupy, use, or alter land or access
    - to divert services
    - to alter water level or flow
    - to restrict public access or navigation
    - to discharge water
    - to fell trees
    - to disturb habitats.

The method of disposal of dredged material may also require land-owner consent, and will require exemption or licensing. This is a responsibility which some Employers may choose to pass to the Contractor and some Contractors' may prefer this if they believe they can offer a more competitive solution than others. However, it should be recognised that the Employer will have more time to investigate and assess various disposal methods than potential Contractors', during the usually short tender period. It should also be more efficient for the Employer to investigate disposal options than requiring all of the tenderers to do so. In addition, the Employer has a duty to see that material is disposed of properly and, if possible, beneficially. These are strong arguments in favour of the Employer deciding and specifying the place and method of disposal, and obtaining all necessary consents and licences, or registering an exemption.

Box 7.5   *Duties of the Contractor*

- To protect the public from harm, especially in respect of working plant, winch wires and ropes, sudden drops, deep water, deep mud or soft soil, and pressure pipes.
- To provide safe and satisfactory working conditions for staff and labour, including provision of fire fighting equipment and training, hard hats, life jackets, life buoys, protective clothing and eye protection, suitable boats for the transfer of persons, suitable arrangements for boarding and landing, and first aid materials and training.
- To contribute to the site safety plan and to regularly liaise with the Safety Supervisor (see Chapter 8).
- To secure fuel supplies and prevent pollution by these or other sources under his or her control.
- To avoid, so far as is possible, extreme variations in water level which might adversely affect the environment, structures, or other to users of the waterbody.
- To avoid, so far as is possible, disturbance to the public in terms of noise, mess, smell or restriction.
- To avoid, so far as is possible, disturbance to the environment, particularly to important habitats, fisheries and aquatic or bank-side flora.
- To avoid unnecessary over-dredging.

Box 7.6   *Matters of joint responsibility*

- Protection of the public.
- Reassurance of the public through information and consultation.
- The safety of the site.
- Compliance with laws and regulations.
- The checking of setting out and dimensions.
- The measurement of the works.

8 Health and safety

This Chapter refers to health and safety issues that arise as a result of maintenance activities which are generally of a smaller scale than capital dredging projects and which may not necessarily be part of the construction process. However, there are a number of regulations that are common to both activities and the overall approach to health and safety issues should be similar.

> **It should be emphasised that health and safety legislation is criminal law and it is strongly recommended that both Client and Contractor organisations seek specialist advice on this topic.**

8.1 STATUTORY INSTRUMENTS

There are a number of regulations which apply to dredging activities, some of which only apply to specific types of dredging.

The Management of Health and Safety at Work Regulations 1992 apply to all work situations and require the identification of significant risks in the dredging process, a strategy to manage those risks and appropriate training of staff to a satisfactory level of competence. In practice the competence and experience of staff will be a key factor in achieving satisfactory working practices.

The Construction (Design and Management) Regulations 1994 (CDM) were introduced to improve consideration of safety issues at the design stage of a project, and to facilitate the effective management of health and safety during the construction process and any subsequent alteration, maintenance and demolition. The Designer's duties under CDM always apply, but other requirements of the regulations will not apply to projects where there will always be less than five people on site at any one time and the construction phase is both for 30 days or less and involves 500 person days or less of construction works. If CDM does apply, the reader is advised to refer to the extensive CDM literature on the application of these regulations.

The Construction (Health, Safety and Welfare) Regulations 1996, which consolidate existing construction health and safety regulations and complement CDM, have replaced the Construction (General Provisions), Construction (Working Places) and Construction (Health and Welfare) Regulations. They deal with the practical construction process and incorporate many general construction workplace requirements as well as those addressing work over or close to water (e.g. the Regulations require that boats used to transport people are of sound construction, properly maintained and in the charge of an experienced boatman; boats must not be overloaded; if there is a risk of drowning then rescue equipment must be kept ready for immediate use and there must be people available who know how to use it). Familiarity with the requirements of these Regulations is essential for those involved in dredging. The Regulations are supported by the HSE guidelines *HSCG ISO Health and Safety in Construction*, 1996.

The Control of Substances Hazardous to Health Regulations 1988 (COSHH) are also significant, especially if contaminated material is to be dredged. They require the Employer to make an assessment of the risks relating to the use of hazardous substances and to include within this the steps needed to comply with the Regulations. Where hazardous substances are known or expected an assessment should establish: the risks to health, the practicability of preventing exposure, the steps needed to control exposure, the likely effects on health, estimates of exposure, and the type and extent of potential harm should the control process fail. Each Employer should evaluate an individual's response to hazardous substances. Compliance with COSHH is achieved through the selection, use and maintenance of suitable controls. These controls may be process controls, or if these are not reasonably practicable, the use of personal protective equipment.

The Construction (Lifting Operations) Regulations 1961, currently under review, require testing and inspection of lifting machinery to ensure its integrity and protection from moving parts. These Regulations will apply to a number of types of dredging equipment. The Reporting of Injuries, Diseases and Dangerous Occurrences Regulations, 1995, require notification of the failure of any lifting machinery to the Health and Safety Executive.

The Provision and Use of Work Equipment Regulations 1992 impose a requirement on the Employer to provide safe and appropriate equipment for the work operations in hand.

The Personal Protective Equipment at Work Regulations 1992 require the Employer to provide appropriate protective and safety equipment for the work operations. In the waterside or over water situations of dredging operations, this would include lifesaving equipment as well as eye, hearing, head, hand and foot protection, and safety harnesses as appropriate. Clothing to protect against the weather must also be provided. However, under the Management of Health and Safety at Work Regulations, 1992, the provision of Personal Protective Equipment is a last resort strategy acceptable only after hazard elimination and mitigation have been attempted.

8.2 RESPONSIBILITIES

The controller of the site is responsible for health and safety issues. Control could be handed over by the Client to the Contractor for the period of the dredging operation, but in many circumstances the Contractor will not be able to isolate a waterway site and so responsibility will often be shared by the Client.

It must be remembered that the Client has specific duties and responsibilities under CDM where those Regulations apply in full (i.e. where there will be more than five people on site at any one time and the construction is for more than 30 days and involves more than 500 person days work). The designer of any works has primary responsibility to inform the Client if such duties exist. Each Employer has duties to his or her Employees for health and safety matters and to others affected by his or her work. Since all health and safety legislation is criminal law detailed consideration must be given to these issues.

8.3 IMPLEMENTATION

The implementation of a health and safety strategy requires a series of staged activities which last **throughout** the life of a project. These requirements follow from CDM, the Construction (HSW) Regulations, and the Management of Health and Safety at Work Regulations, amongst others. They are listed below as a series of actions and responsibilities which effectively provide an audit trail for health and safety assessment:

1. Responsibility for the construction site lies with the controller of the site. In principle this means that the owner (or responsible authority) can hand over control to the Contractor. However, in most dredging works where the site cannot be isolated or closed off, responsibility will, in effect, be shared.

2. If CDM applies then a Planning Supervisor **must** be appointed by the owner (or responsible authority). This may be either someone from within his or her own staff or a suitably qualified external consultant. The Planning Supervisor has extensive duties under health and safety law and his or her first task is to inform the Health and Safety Executive of the project.

3. If CDM applies, then a Safety Plan **must** be prepared before the beginning of construction work, i.e. at the design stage. The purpose of this plan is to pass on information from designers to contractors on matters relating to health and safety (see CIRIA, 1995).

4. A Safety File must be established at the start of the contract and maintained throughout the life of the project, where CDM applies.

5. Following appointment, where CDM applies, the Contractor must develop a construction phase Safety Plan. This will require:

    - identification of all risks associated with the planned activities
    - management and minimisation of the identified risks
    - identification of appropriate training for staff in light of the risks identified
    - identification of particular staff requirements.

    This risk identification, assessment and minimisation phase is key to improving site safety and should be entrusted to staff or specialists intimately familiar with both the legislation and the nature of the activities on a dredging site.

    Where CDM does not apply, the Management of Health and Safety at Work Regulations, 1992, still require all employers and self-employed persons to make a risk assessment. Any Contractor who employs five or more people must record this and develop a safety plan. It is therefore recommended that method statements are produced which set out a safe system of working for all apart from routine activities.

6. On a dredging site, all requirements of the legislation defined in Section 8.1 will need to be taken account of, particularly those relating to lifting operations, working over water and protective equipment. However, all sites are different and general guidance may not necessarily apply, consequently all sites should be assessed independently.

7. The collapse or the overturning or failure of any load bearing part of any lifting machine is a notifiable dangerous occurrence, and must be reported to the Health and Safety Executive as quickly as possible. A written report must also be submitted within 10 days.

8.4 TYPICAL HAZARDS

Dredging operations involve the use of heavy and powerful machinery on or close to water. There is, therefore, a constant attendant risk of accidents causing injury to people or damage to property. In isolated situations it will be prudent to provide the site supervisor with a mobile phone or radio. Appropriate protective equipment should be worn by **all** site staff and visitors.

Floating plant

Before being left unattended, floating plant should be thoroughly checked to ensure that all submerged valves and openings are properly closed and secured. Deck hatches and other potential points of entry by rain or wash water should be checked and secured. All moorings should be carefully placed and secured, particularly in waters subject to variations in level. If spuds are fitted, it is important that they are free to adjust to vertical movement. A list of points to be checked, as illustrated below, should be prepared and displayed on board.

Before leaving plant unmanned, check and secure:

- moorings
- hatches
- deck apertures
- winches
- spuds
- ladder
- submerged intake valves
- submerged joints (such as internal suction pump intake)
- secure heavy deck loads, or machinery
- crane rotation
- crane tracks
- electrics (fire precaution).

Periodically check:

- conditions of mooring and lifting wires and ropes
- fire extinguishers
- life buoys
- non-slip deck coatings
- hatch safety catches.

Floating plant is also associated with the risks of drowning and isolation. Life jackets should be worn by operatives and visitors when on board, or when boarding or landing. Boarding and landing should be at a safe and secure point. If no suitable pontoon or quay is convenient, temporary provision should be made. Life buoys with heaving lines should be fitted to the vessel. Immobile floating plant should not be isolated when manned, i.e. a safety boat suitable for conveying crew to the bank or shore in an emergency should be provided unless the bank is easily accessible.

Land-based plant

The route to be traversed by land-based dredging plant when gaining access or when working should be checked for stability and overhead or buried services, particularly electricity.

If an accident occurs where a machine sinks into soft ground, or falls from a bank (see Figure 2.2) and is immobilised, the first priority is to secure the safety of the operator. If the operator is injured or trapped, the emergency services should be called immediately. When the operator and any other workers involved are safe the recovery of the machine should be planned in two stages.

Stage one is to prevent, or attempt to prevent, any deterioration of the situation. For example, if a machine has slipped part way over a bank and is suspended, further movement should be arrested if possible. Re-entry of the machine by the operator should be avoided unless the risk is clearly minimal.

Stage two - the recovery - may require excavation, blocking, lifting, winching or towing. All such operations may involve sudden movement and the generation of unpredictable forces. The recovery should, therefore, be carefully planned and supervised by an experienced person.

In planning the recovery, the following points should be considered:

- is immediate action essential (i.e. due to rising water level)?
- if immediate action is not essential, await assistance
- what is likely to be the direction and effect of unplanned sudden movement?
- how can sudden movement, or deterioration be prevented?
- what force must be applied to assist recovery, in what direction and by what means?
- if force is applied by lifting, towing, or winching, what will be the effect of failure of the lifting rope or attachment?
- might forces applied in recovery cause further damage?
- what actions are required to prevent spillage of oils and fuels?

9 Environmental considerations

9.1 INTRODUCTION

The water environment supports a great variety of wildlife in the varied habitats of pools, riffles, backwaters, waterside vegetation and bank sides. Its often linear nature also provides a wildlife corridor through the countryside. Historically, dredging of waterways has often had a major impact on these habitats and the wildlife that is supported. There are many examples of poor practice which have left reaches of rivers and canals in a condition of little environmental value (Figure 9.1).

Figure 9.1   Barren artificial channel

Whilst dredged environments may be viewed as artificial they can often be of value and interest and require protection and management. Similarly, areas which have been severely damaged can be managed so that the habitat improves over a period of time. It is now widely acknowledged that a certain level of dredging may be beneficial to many inland waterway environments and indeed is considered to be <u>necessary for beneficial environmental management</u>. However, the approach to dredging must be environmentally sympathetic if severe damage is to be avoided. One of the key principles should be <u>the creation and maintenance of as many different types of habitats as possible which will encourage a greater diversity of plants and animals</u>.

The modification of waterways for flood alleviation, navigation and drainage has a major impact on the nature of the river and its habitats. These impacts vary according to the activity and a number of major consequences are listed in Box 9.1.

Box 9.1 *Major environmental consequences of dredging*

•	River straightening	Increased gradient leading to erosion upstream and sedimentation downstream. Change of bed habitat. Loss of habitat.
•	Channel enlargement	Change of bed and bank side habitats. Frequent installation of artificial lining. Reduced low flow depths.
•	Bank protection	Installation of artificial fringe areas. Change of bed and bank side habitats.

The plant life within the water environment may be of importance in itself, but it also is of great value as shelter and food for a number of fish, invertebrate and bird species. In addition, they provide the necessary environment for the reproduction of some species and can protect against bank erosion. There is an obvious potential for conflict with navigational and flood defence requirements if the plant life is allowed to spread in an uncontrolled way. However, it should also be recognised that increased velocities during high discharge events and vessel bow waves can severely damage plant life by disturbance of the root system, especially in unconsolidated fine grained sediment.

Interestingly, drainage ditches, which are perhaps the most managed waterway environment, are particularly rich in plant life, supporting over three-quarters of the species identified in the UK. A single ditch may contain up to 100 different species, some of which are seldom found elsewhere. Ditches are also important for invertebrate species, including a great variety of molluscs and water beetles. In many places these drainage ditches are the last remnants of extensive wetlands whose once common flora and fauna have become increasingly rare. Guidance to dredging practices in such waterways is provided in MAFF report, *Conservation guidelines for drainage authorities*.

It is also important to recognise the public perception of plant life. Attitude surveys indicate that the public's appreciation of inland waterways is significantly enhanced in those reaches where plant life, especially lilies, are present. Interestingly, publicity material from commercial organisations based on the waterways invariably displays photographic material highlighting plant growth, rather than stark, growth-free water bodies.

This chapter is intended to provide general guidance on the environmental legislation that applies to dredging operations, a method of making relative environmental assessments of alternative dredging methods, planning of dredging operations, monitoring and training. Specific documents dealing with particular issues are referenced in the text.

9.2 CONTROLLING LEGISLATION

Maintenance dredging in inland waterways is subject to limited legislation. In general, capital dredging will be subject to a whole range of regulations whilst maintenance dredging will only be subject to the Wildlife and Countryside Act 1981 and the Water Resources Act 1991. The prime statutory requirements of relevance to dredging in England and Wales are listed below.

1. The Water Resources Act 1991 (Sections 16, 17 and 18) imposes environmental duties on the Environment Agency when undertaking flood defence works affecting water courses.

2. The Land Drainage Act 1994 gives similar duties to the Environment Agency and Internal Drainage Boards with respect to land drainage operations - the duty to further conservation throughout all operations and regulatory activities.

3. The 1994 EC Directive on the Conservation of Wildlife Habitats and of Wild Fauna and Flora (92/43/EEC): Habitat Directive will require that activities affecting certain threatened species of plant or occurring near SACs (Special Areas of Conservation) or SPAs (Special Protection Areas) will require prior consultation with English Nature and probably an environmental impact assessment. Statutory Instrument 2716 is the UK Government's implementation of the Directive.

4. Under the 1981 Wildlife and Countryside Act activities in SSSIs or threatening certain species (plants, animals and birds) must be preceded by written notification to and consultation with English Nature, who have the powers to deny permission for unsuitable dredging activities. English Nature will also give advice on which species are protected by this legislation. A number of lengths of canal are designated as SSSIs and English Nature have initiated the process of consultation as a precursor to the designation of a number of complete rivers. In the future, a number of lengths of inland waterways are to be identified as worthy of special efforts to restore their environmental value.

5. In addition there is an intention by English Nature to increase control in areas adjacent to designated SSSIs, particularly in lengths of waterway upstream of SSSIs which will inevitably have a direct influence on the environment within the designated area.

6. The 1978 EC Directive on the quality of fresh waters needing protection or improvements in order to support fish life.

7. The Land Drainage Improvement Works (Assessment of Environmental Effects) Regulations 1988 (SI1217) is a statutory instrument applying to drainage authorities, which states that where environmental effects are significant an environmental statement must be produced and submitted to the Ministry of Agriculture Fisheries and Food.

8. The British Waterways Act 1995 imposes environmental duties on British Waterways which can be achieved by using their Environmental Code of Practice.

9. Town and Country Planning (Assessment of Environmental Effects) Regulations 1988 (SI 1199) is a statutory instrument that applies to projects requiring planning permission. The local planning authority will decide whether an Environmental Statement is to be produced. Where planning permission is required for particular activities involving dredging, the Environment Agency have an advisory role to the planning authorities and should be consulted if there is a potential impact on water quality or drainage.

10. The 1994 EC Directive on the Conservation of Wild Birds (79/409/EEC): Birds Directive (and RAMSAR Convention) requires the UK to safeguard the habitats of migratory birds and certain particularly threatened species.

11. The UK Biodiversity Action Plan 1995 gives targets and direction for certain key species and habitats, several of which relate directly to the aquatic environment.

12. The Environment Act 1995 requires that wide consultation is undertaken before any works are carried out. The general duty to promote conservation is also included.

13. The Environmental Protection Act 1990 - Part 2, Works of Land, requires that the transport of dredged material, plant and vegetation to a licensed waste site is by a registered carrier. It also imposes a duty of care and the requirement for separate procedures for contaminated waste.

14. Sites designated as Historic Monuments fall under the auspices of English Heritage (in England), who are mandatory consultees for activities in such sites. The Country Sites and Monuments Registers should also be consulted to identify archaeological sites that may affect dredging operations. The responsible authorities in Wales and Northern Ireland are Welsh Historic Monuments and the the Environment and Heritage Service respectively.

15. Consent to discharge into contracted waters are issued by the Environment Agency and are a mandatory requirement; this includes discharges from sedimentation lagoons.

16. A number of environmental designations nominated by national, regional or local bodies and Wildlife Trusts may also exist in waterway environments. These may be based on a variety of particular interests, but do not impose any statutory obligations. However, it is prudent to identify such designations; this can usually be achieved through contact with the local planning authorities.

It should be noted that other legislation applying to related issues, such as health and safety, exists and in this case is addressed in Chapter 8.

9.3 OPERATIONS PLANNING

The planning and management of dredging operations can significantly reduce adverse environmental effects and lead to significant environmental benefit.

Environmental interests may be protected at three stages within the planning of a dredging operation (see Chapter 3):

- establishing a necessity for dredging
- restricting the environmental impact of the works
- creating opportunities for environmental enhancement.

The move towards a requirement to demonstrate the need for dredging and the objective of the works should be accompanied by the establishment of conservation objectives and requirements at the planning stage. It has become clear that a close working relationship between environmentalists, planners and operational staff at an early stage in the dredging planning process has significant benefit to conservation objectives. The principal steps within the planning process relating to environmental issues are described below and are addressed in greater detail in the *New rivers and wildlife handbook* (1994) prepared by the RSPB, NRA and Royal Society for Nature Conservation (RSNC).

9.3.1 Identification of key features

Early in the planning stage of a project the factors of environmental importance should be identified for later consideration. The process should include:

- identification of environmental designations
- consultation with the Environment Agency, English Nature, RSPB, county council(s), local authority(s), Wildlife Trusts and interest groups as appropriate (see Section 3.6)
- scoping study to identify sensitive species in the affected area and the significance of the scheme in terms of its potential impact and hence to determine the need for any further studies
- identification and review of any available data describing fauna and flora
- commissioning of environmental baseline surveys, if appropriate, to identify important species and habitat features in the reaches to be dredged and to establish any areas which should and can be left undamaged; surveys for plants and insects of particular interest should be undertaken in the summer months prior to the planned dredge period since they will be most obvious at this time; adequate time should be allowed for survey planning and resource allocation
- identification of areas of excessive plant life of interest which can be transferred to other reaches
- identification of planning procedure and any requirement for formal Environmental Assessment.

The objective of this identification phase is to provide the basis for the selection of the dredging method and for the planning of operations.

9.3.2 Selection of dredging method

The various alternative types of dredging equipment may have different impacts on the water environment, partly because of the way in which they operate and partly because of their degree of flexibility of use. In general, floating plant is to be preferred because of its reduced impact on the bank during the dredging operation. The relative merits of each method, in terms of the water environment, are addressed in Chapter 5 (Dredging techniques), but a number of other environmental impacts of the plant require consideration:

- noise (for both human and wildlife disturbance)
- bank disturbance
- atmospheric discharges
- oil and fuel leakage into the watercourse
- disposal route impacts
- access damage.

There is a view held by some that the choice of equipment used is largely irrelevant, since their impact is so dramatic that relatively small variations in performance are insignificant. It is more important that the operator is properly trained and instructed in ecological welfare so that suitable cross-sectional shapes and profiles are created and intermittent patches of plant life retained.

A significant factor in plant selection could be the relative ease of creating a good cross-sectional shape with bank-side equipment compared with floating plant. A further factor worthy of consideration is the potentially significant impact on plant growth and bank side stability of traffic levels from the use of barges for repetitive trips to the disposal site, which in turn can cause appreciable damage to plant life.

Whichever dredging method is to be adopted it should be recognised that the disturbance of silts from the bed may result in the suspension of anaerobic material into the water column which is then transported downstream. This is likely to cause oxygen depletion and, in some circumstances, the release of toxic sulphides, methane and ammonia and hence severe damage to many species, especially during the summer months or in areas which are important for fisheries. Where appropriate, the effects of this can be minimised by establishing temporary dam structures downstream or by restricting the flow entering the reach to be dredged.

Relative impact assessment

If a scoping study demonstrates that the environment requires protection then an assessment of the relative impact on the environment of different methods of dredging and disposal is necessary. This is inevitably highly subjective and site specific. It is likely that each individual and pressure group will have a different view. For this reason, it is necessary to attempt meaningful comparison of methods if the optimum method is to be identified.

The approach proposed here addresses the dredging operation only and is numerically based. Each aspect of an operation is assigned a relative score from 0 to 10 whereby a high and negative score denotes high adverse impact. The use of numerical scoring is controversial. Other methods may adopt relative terms, such as good, average and bad, but these remain subjective, and do not permit the same range of impact to be assessed as a 10-point scoring system. The wider range of score provides greater sensitivity.

The objective of the impact assessment is to draw attention to the environmental issues and make those involved consider alternatives and the impacts of their proposals. More important than the particular method used is the quality of the assessment. As a minimum, input should be by an engineer with a thorough understanding of the competing methods, and a scientist with a thorough understanding of the effects or the environment of change wrought by these methods.

For example, an experienced engineer may judge that at a particular site a grab dredger will cause a greater increase in suspended solids than a suction dredger. He or she may be able to estimate the maximum level of suspension in normal operation. The scientist should then judge the impact of the estimated level and duration of suspension on aquatic life. The assessors might then assign respective scores of say, 3 for suction methods and 6 for grab methods. The assessment continues with a similar procedure being applied to other aspects of the proposed operations. On conclusion, the scores for competing methods are totalled and compared.

In the final overall assessment, account must also be taken of practicality and cost. It is sometimes possible to assign numerical scores to these, but if so, the scores must be weighted. For example, if there are to be 10 categories of environmental assessment, each scoring up to 10, the factors of practicality and cost must have greater weight than a single environmental aspect. Whether or not practicality and cost can be represented numerically is very project specific and no guidance can be given here. In general, it is probably better to treat environmental impact as a separate issue.

A proposed form of assessment for the impact of dredging operations is illustrated by examples in Tables 9.1 and 9.2, which have been completed subjectively and in a generalised way. These compare a dragline and suction dredger to be used in a fictional canal maintenance operation. It is important to recognise that assessment of the same methods, but for a different site or application, may result in different numbers. The value of a scoring system is that it should provide a framework for consideration, debate and judgement by professionals of each aspect of potential impact on the environment. The topics listed for consideration in Tables 9.1 and 9.2 are not exhaustive and should be expanded or modified to suit particular site conditions. For a real application the particular characteristics of the site need to be considered as does impact on rare or protected species, locally scarce habitats, and so on.

The basis of scoring can best be explained by use of the example of a 22RB dragline, as shown in Table 9.1.

Turbidity

The first item to be scored is waterway turbidity, which has been given a score of -7. Scores may range from 0 to -10, with increasing negative scores being indicative of greater expected adverse impact. Hence, by assigning a score of -7 significant impact is foreseen, but only in the short term. The method of working of a dragline, whereby the bucket is dragged a substantial distance across the bed during loading, with some material spilling in the process, is bound to force sediments into suspension in the immediate area of working for a short time. Purely in terms of turbidity in a canal, there may be no long-term benefit which is directly attributable to dredging, hence a score of 0, nor is any long-term turbidity likely to persist due to the dredging, hence a score of 0 in the final column.

Bed quality

It is likely that the removal of sediment will be of long-term benefit to bed quality due to the removal of what is probably a highly organic deposit, but the bottom finish will be irregular, which is not desirable in a canal, hence a positive score of 6. The example in Table 9.2, for a suction dredger, anticipates a smooth finish and hence receives a higher score of 8. In the short term, bed quality may suffer, but only very locally. In the long term it will not suffer. Hence a score of 0 in each case.

Marginal plants

By its method of operation the dragline will tend to destroy, or seriously disturb, the working bank margin where reed bed or other vegetation exists. There will be no compensating benefit at the margin, hence a positive score of 0. In the short-term the margin is likely to suffer significant damage, hence a negative score of -7. The margin may take a considerable time, at least one growing season, perhaps longer, to recover. Hence a long-term adverse score of -4.

Other aspects

The assessment and scoring is continued for each category of potential effect. Other effects may be added or deleted. The reader may not agree with each score as shown, but the objective is that two or more assessors, each with a particular field of knowledge, debate the issues and mutually agree a score. On completion, one or more methods will emerge as the preferred method from the environmental viewpoint.

Table 9.1 *Relative impact assessment - Example 1*

Project Proposed method	Example District Canal Maintenance Bank-side 22RB Dragline			
			Relative impact score	
	Influence on	Positive long term	Negative short term	Negative long term
Dredging impact				
Water way	Turbidity	0	-7	0
	Bed quality	6	0	0
	Margins	0	-7	-4
Banks	Wash	0	0	0
	Surcharge	0	-7	0
	Trees and shrubs	0	-9	-7
	Vegetation	0	-6	-1
Agriculture	Crops	0	-10	0
	Stock	0	0	0
Noise	Bankside	0	-4	0
	Locality	0	-2	0
Recreation	Boating	10	-1	0
	Fishing	6	-2	0
	Sailing	0	0	0
	Diving	0	0	0
	General amenity	0	-2	0
Totals		22	-57	-12
Disposal impact				
Transport method	Hopper barge	0	0	0
	Pipeline	0	-6	0
	Side cast	2	0	0
	Bulldozer	0	0	-2
	Dumper	0	0	0
	Truck	0	-7	0
	Crops	2	0	0
	Grazing	0	0	0
	Creation	0	0	0
	Destruction	0	0	0
	Ground	0	0	0
	Surface	5	-3	0
	Safety	0	-3	0
	Amenity	0	-5	-1
Totals		9	-24	-3
Overall totals		31	-81	-15

Note: Categories of influence should be added or removed according to the characteristics of individual sites.

Table 9.2 *Relative impact assessment - Example 2*

Project Proposed method	Example District Canal Maintenance Cutter Suction Auger Dredger		Relative impact score	
	Influence on	Positive long term	Negative short term	Negative long term
Dredging impact				
Water way	Turbidity	0	-2	0
	Bed quality	8	0	-1
	Margins	0	-2	0
Banks	Wash	0	0	0
	Surcharge	0	0	0
	Trees and shrubs	0	0	0
	Vegetation	0	-2	0
Agriculture	Crops	0	-2	0
	Stock	0	0	0
Noise	Bankside	0	-3	0
	Locality	0	-2	0
Recreation	Boating	10	-5	0
	Fishing	6	-2	0
	Sailing	0	0	0
	Diving	0	0	0
	General amenity	0	-1	0
Totals		24	-21	-1
Disposal impact				
Transport method	Hopper barge	0	0	0
	Pipeline	0	-1	0
	Side cast	0	0	0
	Bulldozer	0	0	0
	Dumper	0	0	0
	Truck	0	0	0
	Crops	0	0	0
	Grazing	0	-2	0
	Creation	0	0	0
	Destruction	0	0	0
	Ground	0	-4	0
	Surface	5	-3	0
	Safety	0	-1	0
	Amenity	0	-3	-3
Totals		5	-14	-3
Overall totals		29	-35	-4

9.3.3 Planning of operations

The way in which maintenance dredging is undertaken will have a major contribution to the environmental impact of a scheme. There are several widely accepted factors which should be given careful consideration in order to minimise damage and to maintain or enhance suitable habitats for the ecological components of the water environment. Variability between sites prevents the specification of definitive rules and these are addressed in detail in other documents, including *The New Rivers and wildlife handbook* (RSPB, 1994) and *Nature conservation and river engineering* (Nature Conservancy

Council, 1983). However, a number of key factors can be identified, some of which will apply to each site.

Cross-section

The shape of the cross-section is the single most important factor in maintaining an acceptable environment in the waterway. The most frequent driving forces for dredging (conveyance for flood defence and depth for navigation) have historically led to the dredging of trapezoidal channels. Similarly, there has been a tendency to straighten river channels to remove or minimise bends which cause navigational difficulties and significantly increase hydraulic resistance. In particularly extreme cases, natural channels have been replaced or artificially lined to provide maximum flow capacity with complete disregard for environmental issues (Figure 9.1).

More recently, the impact of such an approach has become apparent and thought has been given to how flood defence and navigational requirements can be achieved whilst maintaining an appropriate regard for environmental constraints. This change in attitude has been accompanied by an increased requirement to consider amenity uses such as fishing, visual amenity and wildlife conservation.

In rivers, an asymmetrical cross-sectional geometry with a steep bank on one side and a more gradually shelving or multistage structure on the offside is preferred, although navigational factors must also be considered (Figure 9.2). The deeper nearside should preferably be dredged down to a firm substrate so that the plant communities have a firm substrate in which to root. The deeper water in this part of the channel also reduces the effect of wave energy, causing less stress to plant communities and providing a better environment for many invertebrate species, whilst at the same time reducing the encroachment of vegetation from the banks.

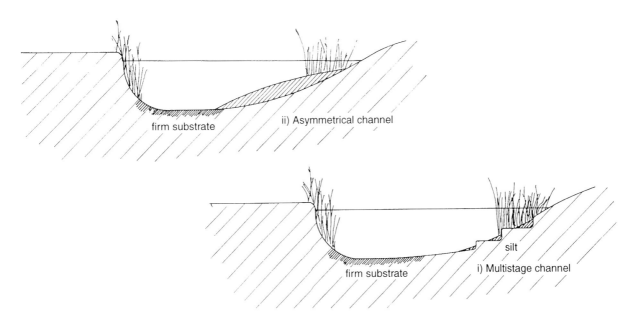

Figure 9.2   *Asymmetrical cross sections*

There is significant environmental benefit in creating or maintaining multi-stage channels which provide waters of different depths across the width. However, the width requirements of this type of channel are considerably greater than for trapezoidal channels, since their hydraulic conveyance is significantly reduced by friction effects in shallow water.

A variation to this approach is the creation of a berm or shallow shelf at the waters edge to provide a shallow water environment. These can be of limited length (a few metres) or extend along the whole length of the waterway, but should only be constructed where appropriate, e.g. along straight banks or the inside edge of meanders.

For environmental reasons it is also advantageous to dredge only short sections, so that general ecological recovery is achieved more rapidly through the migration of seed and species from adjacent undisturbed areas, although this is often not the most cost-efficient approach (see Figure 9.2). A compromise may be to dredge only one bank and to leave a substantial period of time (years) before dredging the other bank. Where possible, a continuous or intermittent margin of vegetation should be left undisturbed at all times at least on one side of the waterway, or for part of the perimeter of enclosed waters.

It is important to recognise that bank erosion may be an important, sometimes the most important, source of sediment causing loss of depth or section. It therefore follows that, where practical, the preservation or creation of a shallow margin of reed bed may provide the dual benefits of reducing siltation rates and enhancing the local environment. Reed margins provide an important refuge for vulnerable species and encourage greater diversity.

Furthermore, to counteract the effects of wash from boats in canals, the establishment or maintainance of a continuous reed fringe is advantageous as this can help stabilise the banks and reduce the suspension of sediment which may cause damage to virtually the whole ecosystem (Figure 9.3). The offside in particular should have a gentle gradient and should be left to be colonised by reeds and other forms of plant life. This growth is important for a number of reasons:

- interest and value in its own right
- wave energy absorption
- absorption of nutrients in run-off
- shelter for a variety of fauna
- food source
- essential habitat for reproduction of some species
- visual amenity.

To enhance such growth there has been a recent trend to place stone material on the bed of some parts of the cross-section of both rivers and canals, to provide a sounder base for rooting and a better habitat for some species of fish and invertebrates.

Figure 9.3   *Retention of reed margin*

Banks

The maintenance of waterway banks is a key activity in environmental conservation. To minimise disturbance and damage on canal and river banks, operations should be undertaken from only one bank wherever possible. Mowing should also be minimised to retain ecological interest and should be undertaken only once each year in the autumn. If possible, some stretches should be left untouched to encourage the growth of taller vegetation which is of particular benefit to birds. The bank profile should be varied to provide a diversity of habitat.

The creation of steep banks in areas where unconsolidated material is not supported by tree roots, requires strengthening in the form of timber, faggots or geotextiles which can be used to provide stability until stabilising vegetation takes over.

The waterside margins (between normal water level and dry areas) are a particularly important habitat and, if possible, should be retained undamaged during dredging. Dredgings placed directly on the banks may have a detrimental effect if they are infertile or if they change the flooding regime of adjacent land.

The visual amenity value of inland waterways is important and most users of waterways are interested in the ecology and appreciate varied vegetation, trees and birdlife. When affecting the bankside, either through management activities, or in order to undertake dredging, it is important to consider the landscape issues of interest to the key user groups (walkers, boating, anglers and cyclists).

The effect of bank-side growth in terms of shade provision is of great importance to the ecology of a waterway (Figure 9.4). Trees provide a number of benefits:

- shade which provides a shelter for fisheries
- several forms of insect life prefer the subdued light of shaded areas
- improved aesthetic appearance of the linear waterway feature
- valuable nesting for several bird species
- roots provide bank stability.

However, trees also result in a high organic load entering the waterway and the necessity for regular lopping of overhanging branches. Trees may also inhibit growth of the types of vegetation which protect against bank erosion in high boat wash locations. Continuous lengths of trees are not desirable for these reasons as well as for aesthetic purposes, the ideal situation being intermittent stands of varied indigenous species accounting for a maximum of 25% of the bank length.

Figure 9.4 *Shaded environment*

Longitudinal profile

As has been noted previously the need for dredging and its extent should be established at an early stage. It is common for localised sediment accretion to occur which, although causing a navigational or flood defence problem, can be dealt with by limited localised dredging. Careful consideration should therefore be given to the required extent and continuity of dredging.

It is widely accepted that intermittent dredging is preferable, since the ecology can then recover by natural recolonisation. This intermittent or sequential approach involves dredging a series of limited lengths of the waterway, whilst the intermittent reaches are dredged in the subsequent years (Figure 9.5). Material being moved downstream in inland waterways shows a great diversity of seed, plant and life forms which will naturally revitalise the ecology of a dredged area.

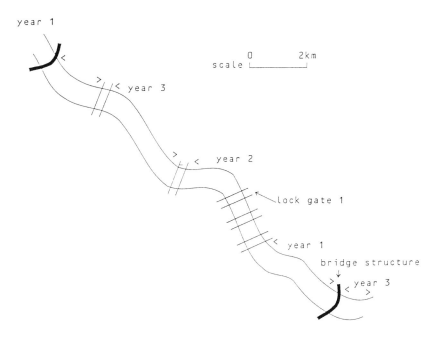

Figure 9.5   *Intermittent dredging programme*

No firm limits on the preferred extent of continuous dredging are yet established, but common sense suggests lengths of 0.5 to 2 km as being practical in economic terms, whilst allowing relatively rapid recolonisation from upstream waters. Very often the limit can be defined by other practical restrictions such as the positions of structures, for example locks and weirs, or arrangements for disposal of the dredged material.

Equally important is the requirement to maintain or create natural variations in river channel flow velocities and depths. In natural river environments shoals, pools and riffles are common (Figure 9.6). Often the effects of the natural features in terms of bed levels is small and their effect on flood events may be limited. However, their environmental benefits are significant and they should be maintained or even introduced where possible. A feasible approach in areas where deepening is necessary, is to over-deepen in places to provide deepwater pools where a harder substrate can be introduced.

Figure 9.6 *River, pools and riffle system*

Seasonal considerations

Dredging should be carried out during the winter months, preferably in late winter as close to the period of natural recovery as possible. The period of least environmental impact from dredging activities is generally considered to be between October and April, which avoids most bird nesting and fish spawning. However, consideration should also be given to migratory fish movements, particularly in rivers where these are of significance. Where no fish spawning sites exist the operational window could be extended to mid May. However, it should be emphasised that the preferred operational window will inevitably be site specific.

It should be bourne in mind that other factors may be of significance when considering seasonality (e.g. boat traffic, freezing, arable crops, etc.).

Frequency of dredging

The frequency of dredging is an important factor in the maintenance of acceptable ecology in inland waterways. Long periods between operations will allow the stronger species to dominate more interesting pioneer species, whilst too frequent dredging will cause undue disturbance and prevent stabilisation of recolonised reaches (see Section 3.4.2).

The preferred frequency is somewhat affected by the level of boat traffic. The objective being to maintain a level of light disturbance so that pioneer species can survive but are not dominated by others which favour more stable conditions. In effect, boats represent 'energy pollution' in the form of wave action leading to suspension of solids and reduced light penetration. If traffic is high, then less frequent dredging operations are desirable; if there is no or very light traffic then, if necessary, dredging every 5 to 10 years may be appropriate.

In ponds and lakes, the frequency is determined by the use of the water. From an environmental viewpoint there are advantages in letting such features evolve into wetlands (i.e. undertake no dredging). However, if the water sustains flora or fauna of particular value, or if fish are of importance, then dredging will be required.

Backwaters and bays

A linear and uninterrupted bank tends to restrict the number of different habitats. It is greatly beneficial to create or maintain backwaters and embayments to provide a wealth of variable habitats (Figure 9.7). These areas can be artificially created, but thought must be given to their form and management. Examples and advice can be found in the RSPB handbook *Gravel pit restoration for wildlife* (Andrews and Kinsman, 1990) and *The new rivers and wildlife handbook* (RSPB, 1994).

Water level

The diversity of aquatic life is dependent on the water depth within the channel and where possible a range of depths should be provided by appropriate cross-sectional shaping. However, variations in water level can be damaging, particularly in periods of low flow and a certain minimum flow and level should be maintained. The draining of individual reaches of canals, or of ponds or lakes for dredging will clearly have a

Figure 9.7 *Example of retained vegetation*

detrimental effect on most species and should be avoided. However, in areas where the environmental attributes are of little interest, such methods may more easily allow environmental management practices designed to restock and recolonise the dredged reach to provide significant environmental improvement.

If de-watering is to occur then guidance relating to recovery described in the following section should be followed.

Where de-watering is planned, local Environment Agency fisheries staff should be contacted. They will undertake fish rescues prior to the operation.

Discharges

Water from dredged material should not be discharged back to the waterway until it is in a satisfactory condition in terms of pollutant load and suspended solids, since turbidity and toxicity can result in fish kills and mortality of other species. Discharges from lagoons will require a consent from the appropriate authority and the Environment Agency should be consulted if the material will travel to a main river.

Recovery

Following a dredging operation recovery of the ecology will not only depend on the nature of the waterway, but also on the sensitivity of the dredging campaign. Recovery occurs primarily through three processes:

> - recolonisation from upstream areas by the transfer of seeds and species into the dredged reach will inevitably occur
> - growth from areas left undredged provides an effective means of speeding recovery (see Figure 9.7) and should be carefully considered; recovery can occur even if the surface silt is disturbed; if dredging in sensitive areas is essential, then many important species can be retained by removing clumps of plant growth before dredging and then returning them to the waterway after completion of the dredging
> - recolonisation from banks and feeder streams.

The speed of recovery will be dependent on the timing of the disturbance. Periods of new growth, reproduction or particularly delicate growth stages should be avoided (see Section 9.3.3 **Seasonal considerations**).

A further factor affecting recovery is the intensity of boat traffic, which causes disturbance and damage to plant life in particular; where possible this should be limited. Activities where high densities occur over short periods, such as rallies or trip boats, should be curtailed during the recovery period.

Recovery of plant life can occur in waterway environments which experience 2,000 boats per year and growth can still be achieved where up to 10,000 boats per year pass if plants are protected from wash.

It should be noted that many important plant species in inland waterways rely on a degree of disturbance and that lengths of conservation interest almost always occur in lightly dredged reaches. If no dredging occurs, the often nutrient rich waters encourage the growth of strong common species which dominate less competitive pioneer species. In disturbed waters the pioneer species are able to colonise and repair most rapidly and it is these that are of particular interest because of their scarcity and because they provide species richness and diversity. Much of the lightly disturbed waterway environment suitable for these species has been destroyed in the UK by development and river 'management' in lowland areas.

In undisturbed nutrient-rich waterways excessive plankton and decay of vegetation reduce dissolved oxygen levels, potentially causing water quality problems for the fauna of the waterway.

Invertebrates are more mobile than plants and recolonise rapidly with downstream flow. This is important as the invertebrate population provides an important food source for fish. As general guidance the existence of healthy vegetation implies satisfactory water quality and levels of disturbance which will provide a suitable environment for a flourishing invertebrate community. However, it is worth noting that most invertebrates will inhabit vegetated areas and generally prefer a hard, coarse bed material rather than soft silts.

Fish populations in canals are often artificial and are dictated by the level of suspended solids, which is usually proportional to the density of boat traffic. The level of suspended solids is strongly influenced by wash interacting with the nature of the channel. Where high suspended solids levels exist, non-visual predators dominate, although species composition may also be affected by stocking practice and by management of the bed. Recovery will therefore depend on management by restocking, control of boat activity and the creation of suitable environments for breeding and food sources.

9.4 MONITORING AND AUDITING

Environmental considerations should be an integral part of any contract and contract measurement method (see Section 7.2). Environmental issues and objectives and operational procedures to achieve them should be formalised as work processes which can be regularly reviewed during the contract. This formal review process will improve Contractor awareness, which will be one of the major factors affecting the level of success of conservation.

Monitoring of the environmental effects of dredging is important for a number of reasons:

- the impact of dredging is in many respects subjective and procedures adopted in each section of the waterway should be evaluated over time to assess whether the adopted approach is as beneficial as possible and to allow knowledge and experience to be gained from each project
- a continuing programme of monitoring allows an audit of the decision-making process which leads to the evolution of the management plan
- by its very existence a monitoring programme will lead to a more thoughtful approach to environmental issues by all concerned in the process
- structured data sets allow monitoring of the condition of the waterway and provide evidence of benefits and commitment to environmental improvement.

There is great advantage in initiating environmental auditing to review the environmental effects of particular dredging operations and management procedures.

This encourages a responsible attitude to the environment by those involved in the process, but also permits the evaluation of decisions intended to promote environmental benefit.

9.5 TRAINING

Appreciation of the importance of environmental issues is the key to successful management of inland waterways. The education and training of those involved in all parts of the dredging process provides significant environmental benefit at project level, but also a more sustained benefit in the long term. Already, the management of many inland waterways has improved immeasurably by including environmental issues in the planning process, by changing the objectives of flood defence works, and by improving work practices by the operations staff.

10 Recommendations

10.1 MONITORING AND DATA ANALYSIS

Monitoring

It is apparent that in the UK understanding of the scale and nature of the past, present and future inland dredging requirement is not adequate. Knowledge is inadequate in respect of the length of waterway requiring regular maintenance, the required frequency of necessary maintenance, the volume and quality of material which must be removed and the character of the waterway and marginal land.

It is recommended that, where absent, appropriate programmes are introduced to monitor and record comprehensively the rate and nature of change within the inland waterway systems. It is expected that this will be of most importance for those canals, lowland rivers and enclosed waters, where the rate of change is most rapid.

The appropriate frequency of monitoring, which should be by site survey, will be site specific, but in most cases is likely to be at intervals in the range 1 to 10 years.

Methods of survey should be reviewed. A range of new survey systems are becoming available which, in navigable waterways, facilitate the rapid collection of comprehensive information concerning bed level, cross-section, condition of submerged structures and features, and bottom sediments. Systems for consideration include:

- swathe sounding
- multi-beam sounding
- interferometry
- ground radar.

A selected system should be used in conjunction with a differential geodetic positioning system (DGPS) and ground modelling software. British Waterways Southern Region have progressed some way along this path.

Dimensional and sedimentary data should be supplemented by progressive field sampling, mapping and desk study to determine the following parameters:

- water quality
- sediment character and quality
- aquatic ecology
- marginal land use
- marginal land ownership
- marginal land ecology
- pollution levels (debris).

A time scale of between 5 and 10 years is recommended for the outline mapping and characterisation of the national waterway systems.

Data analysis

It is recommended that as a minimum, basic geographical information systems (GIS) be established to collate and provide ready access to information arising from monitoring programmes. Compatibility between the systems employed by different organisations is highly desirable, with a common system being preferable.

10.2 CONSULTATION

It is recommended that an appropriate programme of consultation be devised and initiated well in advance of any substantial or sensitive dredging works. Routine consultation should be held with statutory and specialist organisations. For large, sensitive, or potentially controversial projects, it is recommended that the general public be informed and consulted by means of local meetings. Experience has shown that such procedures are helpful in minimising opposition and smoothing the progress of construction.

10.3 FORM OF CONTRACT

It is recommended that a nationally recognised standard form of contract which is appropriate for the proposed works under consideration is used.

The Institution of Civil Engineers publishes a number of standard forms of contract which are well tested and are familiar and acceptable to both sides of the industry. A brief summary of the pros and cons of each follows. In each case, it is strongly recommended that modification of the standard conditions be avoided, or kept to a minimum. Modification may be overlooked by tenderers and the detail of modified or additional clauses may unintentionally influence and alter the meaning of standard clauses, and will not be well tested in law.

Guidance in matters concerning surveys, investigations, specifications and forms of contract for dredging works may be found by reference to BS 6349, Part 5 and Bray, *et al.* (1996).

Specialist advice is strongly recommended when drafting contract terms and when specifying for substantial or complex works of dredging, disposal and treatment.

For dredging and associated construction activities

ICE 5th Edition - first published in June 1973 and revised in January 1979. Well established tried and tested and thoroughly familiar to the majority in the industry. May appear lengthy and cumbersome for works of modest scale, but provides greater flexibility than the 'cut down' version for Minor Works, described below.

ICE 6th Edition - natural successor to 5th Edition. First published in January 1991. Now tending to replace 5th Edition in common use, but in general change has been slow due to the familiarity of most users with the 5th Edition and the fact that some aspects of the 6th Edition are not yet fully tested in case history.

ICE Minor Works - intended to provide a simpler form which is more appropriate to minor works than the 5th and 6th Editions, and less intimidating to inexperienced users. Whilst apparently satisfactory, the Minor Works form does not offer equal scope and flexibility relative to the 5th and 6th Editions, which continue to be used for minor works by many organisations.

For ground investigation

ICE Conditions of Contract for Ground Investigation are a specially targeted set of conditions, published in October 1983, for what is a highly specialised activity. The form of contract is similar to that of the 5th Edition, which can be used as an alternative.

Guidance to the drafting of a specification for ground investigation work may be obtained from *Specification for ground investigation* published by the ICE and Thomas Telford Services Ltd in 1993.

For substantial survey work

ICE 5th, 6th, or Minor Works are all suitable, but may be unfamiliar to some survey companies, particularly if small. This need not preclude their use, but where the scale of survey work is modest, or the competing survey firms are small, award of a contract based upon a simple, but well considered exchange of letters, may suffice.

Guidance to the drafting of a specification for hydrographic work may be obtained from *Guidelines for the preparation of hydrographic surveys specifications* (ICE/RICS, 1983).

It may help both sides of the industry if basic specifications with nationwide application are developed to cover the following broad categories of work:

- dredging man-made drainage channels with land-based plant
- dredging natural rivers with land-based plant
- dredging narrow canals with land-based plant
- dredging canals and navigable rivers with floating plants
- dredging ponds and lakes.

10.4 SELECTION OF CONTRACTOR

It is recommended that tenders be invited only from companies who can demonstrate through good performance on similar works that they are competent to carry out the work required (see Section 7.5). A procedure should be established by which small contractors, or new-comers to the business of inland dredging, may progressively qualify for larger works. Many public authorities will already operate such a system, but may not easily identify suitably experienced contractors for inland dredging. This may be remedied by consultation with other authorities, or specialist consultants with appropriate experience.

10.5 METHOD OF MEASUREMENT

It is recommended that where appropriate the Civil Engineering Standard Method of Measurement (CESMM) is used. However, this will not be the most appropriate or convenient method for all types of dredging work and the associated operations of disposal and treatment. Further guidance is provided in Chapter 7.

It may help both sides of the industry if basic methods of measurement for nationwide application are developed to cover the broad categories of work listed in Section 10.3.

10.6 SCOPE FOR IMPROVEMENT IN LAND-BASED METHODS

Land-based methods of dredging are developing satisfactorily under the initiative of private enterprise and there are no obvious areas where direct investment by public authorities in research is justified.

10.7 SCOPE FOR IMPROVEMENT IN FLOATING METHODS

Methods of dredging conducted afloat have greater scope for development. The ultimate inland maintenance dredging system would have the ability to dredge to a profile of between 5 m and 8 m wide in a single pass with continuous linear progression, producing a final dredged product with natural, or reduced water content, discharged by pipeline, or via conveyor, to shore or hopper. The system would simultaneously remove debris or vegetation for discharge to a separate container or to bank. Such a system does not exist, but in the opinion of the authors, is well within the capability of available technology.

The so called EV dredger, a prototype, which advances linearly with a cut width of 3.5 m and has achieved concentration rates of 90% *in situ* density, already meets some of the ideal objectives, but requires further development. Other systems have been developed by Dutch and Belgium contractors and shipbuilders which also achieve high concentration rates. These are mainly suction type dredgers, but the bucket dredger principle would warrant consideration if development is proposed.

Floating dredgers are most applicable to canals and navigable rivers and hence any development would not find application in all waterway maintenance. The results of analysis in Chapter 2, which should be treated with caution, indicate that the annual volume of maintenance dredging by floating dredgers is about 150,000 m^3, with a value in 1995 of about £1.6 million. These figures exclude the Manchester Ship Canal Company and exceptional dredging at Barton Broad, which are special cases.

Greater incentive for research and development by private enterprise will be provided if substantial contracts for particular types of work are let as long-term contracts of not less than 3 years' and preferably 5 years' duration. It is likely that the maximum benefit will be gained by letting term contracts for work of a type which is recognised to be difficult. The Contractor will then have both incentive and time to develop novel and/or improved methods.

10.8 FUTURE RESEARCH NEEDS

It is considered that direct investment in research in the following areas is most likely to provide real environmental or economic benefits:

- monitoring (see Section 10.1)
- data storage, analysis and access (see Section 10.1)
- standard nation wide procedure and method for categorising environmental characteristics of inland waterway sites
- site measurement of debris (see Section 2.8 - Holland)
- site measurement of contamination (see Section 2.8 - Holland)
- modification of dredged materials for agricultural applications (see Section 2.8 - Belgium)
- procedures for monitoring and recording the impact of dredging operations
- treatment, i.e. research into the state-of-the-art and the reporting of its results, especially 'natural' methods.

It is not recommended that research work already underway in other countries be replicated, but rather that the progress of that work be monitored with a view to adoption of the final product or method where appropriate, or that collaboration and supporting research is undertaken.

Appendix - Pumping

A.1 PUMP HEAD

To pump a soil water mixture through a pipeline, energy must be expended to achieve movement and overcome friction. The laws governing the hydro-transport of soil water mixtures are complex and beyond the scope of this report. However, simple equations may be used to obtain a first approximation of the main factors and to assess the general feasibility of specific proposals. Reference should be made to specialist publications (e.g. BHRA, 1979; Wilson, 1992), or consultants, if the evaluation of complex or large-scale pumping operations is planned.

For a mixture to flow through a pipeline it is necessary to overcome static and friction head. Static head comprises the sum of suction depth below water level and discharge elevation above water level. Friction head when pumping water can be calculated using Darcy's equation (1).

$$\text{Friction head (in metres)} \quad h_f = \frac{0.5.f.L.v^2}{d.g} \quad \quad (1)$$

where:

$f =$ friction factor, see Table A.1

$L =$ total pipeline length in metres

$v =$ velocity of flow in pipe in metres per second

$d =$ pipe internal diameter in metres

$g =$ force of gravity = 9.81.

Equation 1 may be used to calculate friction head in an appropriate length of straight pipe. Further head loss will occur at bends and valves. It is convenient to represent bends and valves as equivalent lengths of straight pipe. The total equivalent length is then added to the straight length and used in equation 1. Typical equivalent lengths for bends and valves of types and diameters most likely to be used for dredging inland waterways are given in Table A.2.

Table A.1 *Typical friction factors for steel pipes*

Pipe internal diameter	Typical friction factor
0.150	0.0154
0.200	0.0152
0.250	0.0150
0.300	0.0149
0.350	0.0147

Note: Steel pipes may become smoother with the prolonged pumping of abrasive fine grain material such as fine sand. In contrast, plastic pipes, which when new may offer lower resistance, tend to become rougher when transporting abrasive materials.

A.2 PUMP POWER

The power necessary to push a mixture through a pipeline is dependent on the rate of flow, the mixture density and the total head. Power can be estimated using the following equation.

Power in horsepower .... $HP = 13.33 \cdot Q \cdot w \cdot H$ ....(2)

where:

$Q =$ rate of flow in cubic metres per second

$w =$ mixture density in tonnes per cubic metre

$H =$ total head = friction head, from (1), plus suction and terminal head.

Table A.2 *Equivalent pipe lengths in metres*

Pipe diameter in mm	150	200	250	300
Bend with radius of 2 x dia.	4	5	7	8
Bend with radius > 3 x dia.	3	4	5	6
Rubber hose radius 10 x dia.	2	2	3	3
90° elbow bend	5	6	8	10
Tee junction	10	13	17	20
Full bore gate valve	3	8	11	16

Greater horsepower than is indicated by equation (2) must be provided by the pump drive engine. This is due to pump inefficiency and transmission losses. Actual inefficiency and losses are dependent on the characteristics of the particular pump and transmission to be used. Further change will result due to pump efficiency varying with the density of the pumped mixture. However, for the purpose of initial assessment, the assumption of a 40% overall combined loss due to pump and transmission will usually provide a conservative answer.

Mixture density is dependent on the ratio of the weight of solids to the weight of transport water. However, in dredging calculations it is more convenient to use the volumetric ratio of bed material to transport water and to convert the volumetric percentage of bed material a weight using the average bulk density of the bed material.

Mixture density = (soil bulk density x Cv) + (density of water x $(1-Cv)$) ..... (3)

where Cv = volumetric percentage of bed material in pumped mixture.

For example, if the bed material has an average bulk density of 1.4 t/m^3 and is pumped at an average volumetric concentration of 15%, taking the density of fresh water as 1.0 t/m^3, the mixture density may be found as follows:

Example: Mixture density = (1.4 x 0.15) + (1.0 x 0.85) = 1.06 t/m^3

Use of the preceding equations requires the estimation of the following:

1. *In situ* bulk density of bed materials - see Table A.3.

2. Velocity of pumped mixture - see Table A.4.

3. Volumetric concentration of bed material in pumped mixture - see Table A.5.

Table A.3 *Typical bulk densities for various hydro-soils*

Soil type	Typical occurrence	Typical density
Highly organic	Lakes and ponds	0.9-1.2
Soft clay, silt	Canals, tidal rivers	1.2-1.4
Sandy silts	Canals, slow rivers	1.4-1.7
Sands and gravels	Fast rivers	1.7-2.0

Table A.4 *Typical velocities appropriate for pumped mixtures*

Soil type	Minimum velocity m/sec	Typical velocity m/sec
Highly organic	0.5	3.0
Soft clay, silt	1.5	3.5
Sandy silts	2.5	3.5
Fine to medium sand	3.5	4.5
Gravel	4.5	6.0

Note: The determination of critical velocity for gravels is complex. Reference to specialist publications or consultants is essential if planning to pump significant volumes over distance in excess of 250 m.

Table A.5 *Typical volumetric concentrations for pumped soils*

Soil type	Average %	Maximum %
Organic	25	100
Soft clay and silt	20	50
Sandy silts	15	30
Fine to medium sand	10	20
Gravel	5	15

Note: The average concentration achieved is highly variable, being dependent on many factors, including material type, thickness to be removed, and type of cutterhead. It should be apparent, therefore, that the above figures must be used with extreme caution. No substantial project which involves pumping should be planned without consultation with in-house specialists, or specialist consultants or contractors.

A.3 SETTLING TIME OF FINE SOIL PARTICLES

The settling time of soil particles may be important when determining the dimensions of bunded lagoons into which water-soil mixtures are to be pumped and stored, or when estimating the distance which suspended sediments may travel before settling to the bed. The settling velocity, sometimes called fall velocity, of soil particles may be estimated using Stokes Law.

Stokes Law, in simplified form, states that the settling velocity (v) of a soil grain is given by:

$$v = \frac{2.g.d^2.(Gs-1)}{9.400.n} \quad \quad (4)$$

where:

v = particle settling velocity in cm per sec

d = particle effective diameter in mm

g = force of gravity = 9.81

Gs = specific gravity of soil particle

n = viscosity of water, at 20°C = 0.010050, at 10°C = 0.013077.

For initial estimates, taking the value of n as 0.01 for water at 20°C and an average value of 2.65 for Gs, this may be simplified to:

$$v = 90.d^2 \quad\quad\quad\quad\quad\quad\quad\quad\quad\quad\quad\quad\quad\quad\quad\quad\quad\quad\quad\quad (5)$$

A range of results are shown in Table A.6.

Table A.6 *Settling velocity for soil particles*

Particle diameter in mm	Settling velocity in cm/sec
0.010	0.009
0.050	0.225
0.100	0.900
0.150	2.025
0.200	3.600
0.300	8.100
0.400	14.400
0.500	22.500

References

Andrews and Kinsman (1990)
Gravel pit restoration for wildlife
Royal Society for the Protection of Birds

Anglian Water Authority (1982)
Conservation guidelines for river engineers
Environment Agency

Anon. (1994)
Environmental concerns breathe new life into the maintenance dredging industry
Dredging and Post Construction, January 1994

BHRA Fluid Engineering (1979)
A guide to slurry pipeline systems
BHRA, Cranfield

Bray R N, Bates A D and Land J M (1996)
Dredging: a handbook for engineers
Arnold, London

BS 6349, Part 5 (1991)
Code of practice for dredging and land reclamation
HMSO, London

BS 7121, Part 1 (1989)
Safe use of cranes
HMSO, London

CEDA/IADC (1996)
Dredging and the environment
CEDA, London

CIRIA (1995)
CDM Regulations - case study guidance for designers, an interim report
CIRIA Report 145, London

CIRIA (1996a)
Guidance on the disposal of dredged material to land
CIRIA Report 157, London

CIRIA (1996b)
Design and management of constructed wetlands for the treatment of wastewater
CIRIA Research Project 507, Funders Report/CP/45, London

Detzner (1993)
New technologies for the treatment of dredged material
Proceedings CATS II Congress 1993

DoE (1994)
Environmental Protection, The Waste Management Licensing Regulations 1994
HMSO, London

DoE (1995)
Development of a National Waste Classification Scheme, Stage 2: a system for classifying wastes, Consultation Draft
HMSO, London

HSE (1996)
Health and Safety in Construction
Heath and Safety Executive, HS(G)150, London

ICE (1979)
Conditions of Contract 5th Edition
Institution of Civil Engineers, London

ICE (1983)
Conditions of Contract for Ground Investigation
Institution of Civil Engineers, London

ICE (1988)
Conditions of Contract for Minor Work
Institution of Civil Engineers, London

ICE (1991)
Conditions of Contract 6th Edition
Institution of Civil Engineers, London

ICE (1993)
Specification for Ground Investigation
Institution of Civil Engineers/Thomas Telford, London

ICE/RICS (1983)
Guidelines for the preparation of hydrographic surveys for dredging
Institution of Civil Engineers, London

Nature Conservancy Council (1983)
Conservation and river engineering
Nature Conservancy Council

Pearson R C and Jones N V (1975)
The effects of dredging operations on the benthic community of a chalk system
Biological Conservation (8)
Applied Science Publishers Ltd, England

PIANC (1990)
Supplement to Bulletin No. 70, *Management of dredged material from inland waterways*
PIANC, Brussels

PIANC (1992)
Beneficial uses of dredged material: a practical guide
PIANC, Brussels

PIANC (1996)
Handling and treatment of contaminated dredged material from ports and inland waterways "CDM"
PIANC, Brussels

RSPB (1994)
The new rivers and wildlife handbook
Royal Society for the Protection of Birds, National Rivers Authority and Royal Society for Nature Conservation

Rijkswaterstaat (1996)
Contaminated sediment remediation
North Sea Directorate, Netherlands

Schotel F M and Rients H (1993)
Chemical treatment and immobilisation of contaminated sediment in DPTP Phase II
CAT's II Conference Proceedings

US Army Corps of Engineers (1985-1996)
Environmental effects of dredging program - technical notes and D-series
Waterways Experiment Station, Vicksburg, USA

Wade P M (1978)
The effect of mechanical excavators on the drainage channel habitat
EWRS (European Weed Research Council) 5th Symposium on Aquatic Weeds

WTC (1995)
Proceedings of Sediment Remediation '95
Waste Water Technology Centre, Canada

Wilson K C (1992)
Slurry transport using centrifugal pumps
Elsiver Applied Science, New York

Inland dredging - guidance on good practice

Inland dredging operations account for millions of pounds of spending in the UK annually, yet waterway practitioners recognise the significant scope for improvement in current dredging practices. Considerable cost savings could be achieved, for example, by improving operational effectiveness and efficiency.

The aim of this initiative was to review current technologies, establish operational and environmental good practice, clarify contractual obligations, and to propose a way forward.

This report includes guidance on good practice with regard to dredging and sediment treatment techniques, including consideration of utility, flexibility, cost and environmental effects, and on the management of dredging operations. It identifies good practice; defines an appropriate approach to project evaluation, task definition and option selection; considers measurement techniques, contractual relationships and relevant health and safety issues; and outlines further research needs.